Occupational Hazards of Pesticide Use

Occupational Hazards of Pesticide Use

Edited by

Graham J. Turnbull

Assistant Editors

Derek M. Sanderson
John L. Bonsall

Taylor & Francis
London and Philadelphia
1985

UK Taylor & Francis Ltd, 4 John St, London WC1N 2ET

USA Taylor & Francis Inc., 242 Cherry St, Philadelphia. PA 19106-1906

British Library Cataloguing in Publication Data

Occupational hazards of pesticide use.

1. Pesticides — Safety measures
I. Turnbull, Graham J. II. Sanderson, Derek M.
III. Bonsall, John L.
363.1'9 SB952.5

ISBN 0-85066-325-3

Library of Congress Cataloguing in Publication Data

Main entry under title:

Occupational hazards of pesticide use.

Includes index.
1. Pesticides — Toxicology — Addresses, essays, lectures. 2. Pesticide applicators (Persons) — Diseases and hygiene — Addresses, essays, lectures. 3. Pesticides — Application — Safety measures — Addresses, essays, lectures. I. Turnbull, Graham J. II. Sanderson, Derek M. III. Bonsall John L. [DNLM: 1. Environmental Exposure. 2. Pesticides. WA 465 01353]
RA1270.P4023 1985 363.1'79 85-12607
ISBN 0-85066-325-3

Printed in Great Britain by Taylor & Francis (Printers) Ltd, Basingstoke, Hants.

<u>CONTENTS</u>

FOREWORD

Dr. R. Murray
President, International Commission on Occupational Health.

The word pesticide is now so much a matter of everyday usage that it is difficult to appreciate that it has only been current for a generation. There have always been pests, in the sense that mankind has always had to coexist, often impotently, with other living organisms, troublesome, annoying, harmful or destructive, which compete with him for food and shelter or act as parasites or agents or vectors of disease, so threatening his life or at least his quality of life. Until World War Two, the armamentarium of substances that could effectively dispose of pests was very limited. Since then, starting with the organochlorine insecticides, there has been an explosion of chemical technology that has produced a wide range of highly specific pesticides. The skill today is in devising a poison which kills the pest while leaving the rest of the environment undamaged. If we cannot yet achieve this utopian ideal, then reliance must be placed on the skill of application, so that the maximum dose of pesticide is directed at the pest and the minimum to the creatures of the environment, either directly or via the food chain. Particular care must be taken to ensure that the operator who applies the pesticide, and is therefore closest to potential danger, is educated, trained and protected.

The problem is worldwide, affecting all countries, and the World Health Organisation and many individual countries have devised schemes to make the best use of pesticides with minimal dangers to operators and to the balance of nature. This book produced by a group of experts in the subject is an invaluable compendium of information on all aspects of pesticides exposure and I commend it wholeheartedly.

Robert Murray
May 1985

PREFACE

"Everything should be made as simple as possible, but not simpler"

Albert Einstein

Pesticides bring undoubted benefits despite their potential capacity to be harmful to man. Use of pesticides is widespread in agriculture and public health and there is a well-recognised need to handle them with common sense in order to minimise exposure and avoid harm. Are these precautions effective and what are the health effects of occupational exposure to pesticides during use?

The hazard of a pesticide in use in agriculture or public health depends upon its toxicity, and upon the way it will be used, especially the degree of exposure of users.

The degree of toxicity is a property of the individual chemical product, and has to be evaluated completely, according to national regulations, before the product is put on the market. The other component of the occupational hazard evaluation, exposure during use, is a function of the job being done, how it is done, and the formulation type, and can be calculated from published data on exposure to another similar product during similar use.

Hitherto, there has not been a systematic scientific approach to such calculations of exposure and hazard during application of pesticides to complement the very full evaluations of safety of pesticide residues in food. This book proposes just such a systematic scientific approach to the realistic assessment of occupational exposure to pesticides during application, and, hence, safety.

Factual information is presented on the occupational exposure of those who use pesticides in agriculture or public health. It gives an insight into how fundamental calculations of hazard may be employed to control the use of pesticides in order to ensure the safety of those occupationally exposed to them.

There is a surprising amount of information on occupational exposure to pesticides. By presenting it concisely, the authors provide access to important information on pesticide safety.

The first part of this book describes how information on occupational exposure can be obtained by field studies and laboratory analyses. It then summarises the state of knowledge according to the nature of the handling tasks with pesticides. Later chapters examine the evidence for pesticide application having harmful effects on health.

Use of pesticides was very limited until, during the last half century, chemists devised an increasing number of pesticidal chemicals. Many of the early pesticides, such as arsenic and nicotine, were highly toxic. Early pesticide application equipment was unsophisticated. Both protective clothing and safety precautions were unrefined and a number of fatalities occurred from the use of pesticides such as parathion or dinitro-ortho-cresol. Many countries then established national regulatory systems to control the use of pesticides and determine the precautions to be taken. As described in Chapter 6, in each country there is control of exposure during application and of any residues of pesticides in food crops or animal products. The national authorities regulate the permitted uses of pesticides and decide on safety precautions. The submissions by manufacturers for registrations are reviewed in detail by the national authorities and must include extensive information on the safety of the pesticides to man and to the environment.

What may have been acceptable working conditions even fifty years ago certainly would not be regarded as acceptable today. Hence, there is a need to consider, in more detail than ever before, the extent of occupational pesticide exposure and to weigh the evidence of ill-effects of that exposure.

This book sets out to satisfy that need.

The editors acknowledge the helpful co-operation of the authors, the extensive support of FBC Limited in providing facilities, the considerable advice from excellent colleagues, and permission to publish information from exposure studies performed by FBC Limited and Ciba-Geigy Limited. Helen Johnson prepared the manuscript with skill and great patience.

Graham Turnbull
Harston
March 1985

CONTRIBUTORS

John L. Bonsall, M.B., B.S., M.R.C.S., L.R.C.P., D.I.H., A.F.D.M.
Company Medical Officer
FBC Limited*
Hauxton
Cambridge

John F. Copplestone, M.F.C.M., D.P.H., D.I.H.
Chief, Pesticide Development and Safe Use Unit, Division of Vector Biology and Control
World Health Organisation
1211 Geneva 27
Switzerland

Steven J. Crome, B.Sc.
Huntingdon Research Centre
Huntingdon
Cambridgeshire

Roy Goulding, B.Sc., M.D., F.R.C.P., F.R.C. Path.
Institute of Industrial and Environmental Health and Safety
University of Surrey
Guildford
Surrey

Derek Sanderson, L.R.S.C., M.I. Inf. Sci.
Environment, Product Safety and Registration Group
FBC Limited*
Chesterford Park Research Station
Saffron Walden
Essex

Ray Tincknell, Ph.D., B.Sc., A.R.C.S., C.Chem., F.R.S.C.
Regulatory Consultant
11 Walkwood End
Beaconsfield
Bucks

Graham J. Turnbull, Ph.D., Dip. Tox.
Environment, Product Safety and Registration Group
FBC Limited*
Chesterford Park Research Station
Saffron Walden
Essex

* A pesticide and industrial chemicals company wholely owned by Schering Aktiengesellschaft, Berlin and Bergkamen (West Germany).

CHAPTER 1

PESTICIDE SAFETY

R. Goulding

INTRODUCTION

The chemicals used in agriculture, horticulture and public health to control the predatory forms of life that interfere with food production, or which transmit disease, all come within the definition of pesticides. By their very nature pesticides are biocidal and so they may present risks to people coming into contact with them and, above all, to workers handling them during use.

The purpose of this book is to evaluate the form and extent of the occupational health hazards likely to arise from pesticides during use and the means by which these hazards can be controlled.

HISTORY

A continuing struggle is waged worldwide against a recurrent inadequacy of food supply and against vector-conveyed disease. In biblical times the tares choked the corn and the locust pestilences obliterated virtually all plant growth in their wake. The population of medieval Europe was decimated by bubonic plague, spread by the many rats. More recently the grape vines of France were almost wiped out by Phylloxera. In this struggle materials which were obviously toxic, such as nicotine, arsenic, strychnine and phenol, were efficacious over the years, but they had their disadvantages. Firstly, they were indiscriminate in their toll, tending to destroy any living organisms, or tissues, with which they came into contact and, secondly, they could imperil the people handling them, especially without due care. The extent of the casualties suffered in this way has never been reliably recorded.

Following the Second World War, a rapidly expanding industry sythesised many completely new chemicals and screened them for pesticidal activity. A wide range of selective pesticides has now been found which will suppress particular fungal pests, harmful insects, nematodes, weeds and rodents. Today, the community has at its command a wide range of efficacious pesticides, for better or for worse.

RISK VERSUS BENEFIT

To those people faced with starvation due to crop failures, or upon whom, say, the malarial parasite had wreaked havoc, these new pesticides have been a salvation. On the other hand, in the more advanced societies, where a high, comfortable and more assured standard of living seems to be the rule, the dangers of pesticides loom more ominously and some people are now questioning whether such aids are needed at all. Ultimately, any decisions taken should pay proper heed to both risks and benefits and, among the risks, those to operators due to exposure while handling the chemicals. It is this aspect with which this book is pre-eminently concerned.

DEFINITIONS

Dictionaries are not very enlightening over the term pesticide. It is a word that is not simply defined. Instead, it needs to be descriptively explained, conveniently in the manner adopted by the Pesticides Safety Precautions Scheme in Great Britain. Thus it includes those products intended, or employed, for the purpose of:-

(i) Destroying organisms harmful to plants, or to plant products (including grass), or protecting plants, or plant products from such organisms, e.g. insecticides;

(ii) Destroying organisms harmful to wood, or protecting wood from such organisms (excluding decorative products applied to the surface of wood and not penetrating it), e.g. fungi, etc., which cause timber decay;

(iii) Improving, or regulating, plant growth and/or development (excluding products which are purely plant nutrients, or soil conditioners), e.g. cereal tillering agents;

(iv) Preserving plant products, but excluding those substances deliberately added to foodstuffs to preserve them, e.g. insecticides and related substances that suppress destructive organisms in stored grain;

(v) Destroying, or inhibiting, the growth of plants, e.g. herbicides;

(vi) Destroying parts of plants, or preventing their undesired growth, e.g. anti-sprouting agents for potatoes;

(vii) Rendering harmless, destroying, or giving protection against any nuisance, or harmful animals, or insect pests, e.g. rodenticides;

(viii) Controlling organisms with harmful, or unwanted, effects on water systems, buildings, or other structures, or manufactured products, e.g. anti-fungal agents on stone, or brick, structures;

(ix) Protecting animals against ectoparasites (excluding products for direct administration to animals) e.g. sheep dips.

Elsewhere, of course, the wording may not be identical, but the meaning is much the same. The term pesticides is comprehensive, if not omnibus, in its scope and embraces a series of sub-divisions based on usage. These include:-

(i) Insecticides, aphicides, nematocides, larvicides, etc., that are directed chiefly against undesirable small and invertebrate forms of life.

(ii) Fungicides, to control fungi and moulds, such as mildews on plants and plant products.

(iii) Herbicides (weed-killers), to abate unwanted plant growth, either broadly or more specifically.

(iv) Rodenticides, intended to poison rats, mice and other vertebrate animal vermin.

(v) Plant growth agents, to restrain, for example, the growth form of bushes, or the tillering of cereals.

(vi) Protective agents for building materials and buildings.

No pesticide, or even single group of pesticides, can possibly cater for all these requirements. The converse principle is today the rule, with separate chemicals being tailored for the task in mind and exerting little influence otherwise. Thus, a fungicide or herbicide should not at the same time be an insecticide. Carrying this rationale still further, the different pesticides have become so selective that, for example, wild oats can be singled out for destruction without adversely affecting the wheat crop and aphids can be controlled without harm to beneficial insect populations.

As a corollary to this concept, the molecular configuration of the various pesticides also exhibits a wide diversity in accordance with the separate purposes envisaged. The specificity of the effects of present pesticides means that, in general, the likelihood of man suffering harm is very much reduced because the multiplicity of chemical families reduces the exposure to each.

PESTICIDE GROUPINGS

Among the earliest and undeniably effective insecticides were the so-called organochlorine compounds, including aldrin, dieldrin, DDT, endrin and chlordane. Each of these active ingredients is a chemical with a defined identity. These pesticides were outstanding in their unusual physico-chemical stability, allied to which property was their biological persistence. This attribute was valuable for insect control but it led to a ubiquitous presence in the environment and, by intervention into the food-chain, to a potentially harmful accumulation in the tissues of certain species, notably those of raptorial habits. For this reason their use in many countries has been almost entirely phased out as alternatives have been invented.

In their place the organophosphorus compounds have come to occupy a prominent position. Characteristic of this group is the ability to interfere with the naturally-occurring and vital cholinesterase enzyme systems of the body. By this mechanism they are insecticidal, but equally they may prove toxic to the human body to which they gain access. Of similar biochemical behaviour are the carbamates, though their duration of action tends to be somewhat shorter.

Then there are the synthetic pyrethroids, modelled upon the insecticide pyrethrin, for which particular species of chrysanthemum serve as a natural source.

Besides these major groupings there is a host of other insecticides also in commercial use.

Turning to the fungicides these, too, exhibit a wide diversity of chemical identity and specificity. These include a number of compounds of which the heavy metals constitute an essential part, such as mercury and tin.

Among the herbicides, or weed-killers, the most widely developed are those based upon chlorophenoxyacetate, with 2,4-dichlorophenoxyacetic acid (2,4-D) and MCPA being popular. These, by the peculiar manner in which they interfere with the metabolic processes of plant growth, act selectively against broad-leaved weeds. Others, by contrast, single out the grasses, or monocotyledons, as their target, whereas the dipyridyls, such as diquat or paraquat, are inimical to all forms of plant life with chlorophyll-dependent photosynthesis. Yet again, for total weed eradication on, say, railway tracks and driveways, the chlorates and the triazines, e.g. simazine, and glyphosate are preferred. Once more, though, there exists a multiplicity of herbicides chemically quite apart from those mentioned above.

For rodent control, the blood anti-coagulants are extensively favoured, above all warfarin. These bring about delayed death. Much more acutely lethal are the fluoroacetates.

With such diversity in their chemical make-up and in their mode of action against pests it is not surprising that these various compounds display no uniform pattern of toxicity, qualitatively and quantitatively, towards man. It is, therefore, scientifically and medically meaningless to allude to pesticide poisoning as an expression having any diagnostic connotation. The individual pesticide, or its generic grouping, should always be stated.

In practice, moreover, another possible complication must never be overlooked, for the active ingredients of pesticide chemicals are seldom applied alone. Customarily they are incorporated into formulations, together with solvents, carriers, wetting agents and the like, any of which may confer upon the mixture an intensification, a reduction, or a modification of the toxicity of the primary constituent. Of further importance is the recognition that the product, as supplied, is commonly in concentrated form and thus is likely to be more noxious than the material that is diluted before it is applied.

PRE-MARKETING TOXICOLOGICAL EVALUATION

To release new, possibly dangerous pesticides on to the market without having formed some clear idea beforehand of their capacity for harm would be cavalier in the extreme. Since deliberate experimentation on man is ethically discountenanced, the accepted and, indeed, obligatory manner by which to arrive at a toxicological evaluation is by scientifically designed and meticulously conducted animal experiments, chiefly with rats and mice, along with guinea pigs and rabbits and occasionally with other species as well. The testing programmes set down by national authorities are both extensive and searching as can be seen from Chapter 6. Analytical techniques have been perfected for measuring pesticides in organs and biological specimens such as blood and urine. The animal studies embrace acute, short-term and long-term toxicity by different routes of administration, tests for skin and eye irritation and sensitisation, carcinogenicity, mutagenicity, possible interference with reproduction and embryonic development and the metabolic fate of the chemical in the body. Other specialised tests may be carried out as appropriate. At the same time the susceptibility of a number of species of wild life to the chemical is also checked, and its fate in the environment - in soil, in air, in water is investigated. Seldom is it possible to complete the major part of this task in under five years, but eventually a toxicological characterisation is assembled. This affords a basis from which to forecast the hazards likely to be encountered by (i) the workers who will be called upon to handle the chemical, (ii) the consumers who may unwittingly ingest pesticide residues lingering within components of their diet and (iii) the environment generally, with special concern for other species.

At this stage it becomes imperative to differentiate between toxicity and hazard. The former reflects the intrinsically noxious capabilities of a substance, as arrived at objectively and mainly by animal experiments. The latter depicts the likelihood of harm, or danger, that may arise from exposure to the substance under defined circumstances. The hazards attending any pesticide, therefore, must be regarded in strict relation to its usage and this is discussed in more detail in Chapter 2. However, it is important to emphasise at an early stage that industry itself rejects many, indeed most, candidate pesticides. Most of the novel compounds synthesised in the research and development laboratories lack the necessary efficacy, or are shown by the toxicological evaluation to have undesirable features, and therefore are never put on sale.

METHODS OF APPLICATION

There are many types of pest and their habits are diverse. Obviously no single and universal application technique will meet all the needs so the equipment used to apply pesticides takes on a variety of designs.

Traditionally and widely adopted is the practice of purchasing a liquid concentrate, diluting it with water on site and then applying this as a spray. Analogous to this are "wettable powders", supplied in the dry state, but prepared for use by being taken up in water in a finely dispersed state, without actually dissolving.

Such-like liquid preparations may be simply applied from a watering-can or, rather more elaborately, by the aid of a hand-directed 'knapsack' ('back-pack') sprayer, from which the droplets are ejected under pressure.

If the same diluted liquid is carried within a tank, either attached to the rear of a tractor, or towed behind as a trailer, with the spray-bar at the back and directed downward - as is the habit where large acreages have to be covered, then so long as the driver remains within an enclosed cab his exposure is virtually nil, except when he is diluting and tank-filling. A variant of this procedure, however, is for the spray under high pressure to be directed upwards (Figure 1.1), or entrained in an air stream (air blast) as when treating orchards of top-fruit. Then the driver is much more liable to become contaminated, unless there is a cab which is well sealed.

To minimise the bulk of liquid to be transported, controlled droplet application (CDA) techniques have been introduced, whereby the equivalent amount of pesticide is applied in a much lower volume (<60 L/ha). If the spray falls upon the driver the amount of active ingredient to which he (or she) is exposed is that much higher. For some purposes the volume applied can be reduced to the minimum and such ultra-low volume (ULV) applications commonly use virtually no diluent, or an oily non-evaporative diluent and rates of less than 5 L/ha.

An alternative to liquid formulations is to have the pesticide presented at low concentrations in the form of fine granules which are disseminated as such in the dry state. Drift is thereby lessened, while contact with the body can be of minor consequence. Pesticide incorporated at a low concentration in a dust may be used to treat seeds before planting and in public health insect control.

For some purposes, the active ingredient is introduced into paints which are used as coatings for walls, hoppers, etc., by brushing-on, or spraying. For wood preservation, timber may be immersed in tanks of chemical.

In the process of fumigation gases may be released directly into closed premises, e.g. glasshouses, silos, containers, etc., or fluid may be intermingled with the commodity and allowed to evaporate. On the domestic scale small devices are available from which the pesticide volatilises spontaneously into the atmosphere.

When it comes to dealing with rodents and rabbits,

FIGURE 1.1 Orchard Spraying With Upward Directed Nozzles

Spray under high pressure penetrates the foliage but also reaches the driver.

noxious gases may be delivered directly into the holes or burrows, though for rats and mice the active principle usually is mixed in baits that are attractive to the animals and upon which they feed.

Each pesticide, and each sort of application equipment, is intended only for certain approved uses. Work practices which cause needless personal contamination, the use of inappropriate equipment, or application on a crop (or at a time in the season) which is not approved, are misuses of the pesticide. Misuse may also include inadequate storage conditions or inappropriate handling during transport. Occupational exposure to pesticides may be increased by their misuse, even to the extent that there is over-exposure and some adverse health effect. Normal use, however, involves a low level of occupational exposure, either by skin contact or inhalation, as described in Chapter 3.

HAZARD ASSESSMENT

From this cursory survey it can be seen that the hazards facing those who work with pesticides cannot be assessed from the toxicity of the pesticide, nor even of the product, alone. As described in Chapter 7, there must also be a consideration of dilution factors, precise mode of usage, possibly weather conditions and the effects of any protective clothing that may be worn during pesticide application.

Protective clothing includes aprons, overalls, gloves, masks, goggles or visors worn in addition to the usual work clothing for protection from contamination. Often the footwear is also chosen to provide similar protection. In contending with the hazards, the practicability and efficiency of protective measures must never be overlooked. Once the major routes of exposure have been identified, then evasive strategy and precautions can often be taken to close them, whether by gloves and so on, or perhaps more crucially, by altering the working methods. The effectiveness of protective clothing is discussed in later chapters in relation to the distribution of exposure over regions of the body.

In spite of the safety precautions, is the chemical likely to fall upon the skin, or get into the eyes? If so, will it be irritant, corrosive, or bland? The animal findings may assist in answering such a question. In this respect the concentrate may behave more vigorously than the diluted material.

Is there a prospect of repeated skin contact and, if so, is there a risk of inducing sensitisation? The findings in the guinea-pig sensitisation study may give a clue here.

Recognising that, occupationally, the possibility for systemic pesticide poisoning resides in the avidity with which the chemical may be taken up by the body percutaneously, how much of it, in the context of the way it is being used, is likely to fall upon the skin? And when it is so deposited, what proportion of it will penetrate the skin barrier? Reliable, quantitative observations are indispensable to the reaching of a sound conclusion, as described in Chapter 7.

Experience has shown that, ordinarily, pesticide application is not responsible for a high ambient concentration of the chemical in respirable form. The spray droplets mostly are too coarse to be inhaled and dust formulations and granules are manufactured and used in ways which minimise very small particles entering the breathing zone. Nevertheless, wherever a high level of toxicity by inhalation has been demonstrated in animals, the chance that workers may inhale the material should always be carefully examined.

For operators, the risk of their becoming poisoned by ingestion is a remote one, unless the chemical has a high oral toxicity and they are inordinately careless about their eating or smoking habits and personal hygiene. The adverse effects which may result from over-exposure to pesticides are discussed in Chapters 4 and 5.

Until it can be established that, under the conditions of use, there is absolutely no exposure and no uptake of pesticide by the workers, regard must be paid to what has been discovered from the animal studies about long-term toxicity, carcinogenicity, mutagenicity, effects on reproduction and so on. For a genotoxic carcinogen, and taking a cautious attitude, it is widely believed that there is no threshold dose below which safety can be predicted, so such chemicals are vetoed completely. For the remainder of the adverse reactions listed (and possible for non-genotoxic carcinogens), it may be feasible to calculate a safety-margin between the highest no-effect-level for that toxic effect found in the animals and the dosage taken up over a normal working day by a person. If this margin is judged to be sufficiently wide for safety, then approval may be granted.

For accuracy to oust a combination of guesswork and inspiration in these hazard predictions, then ever more emphasis must be placed upon the quantitative determination of just what exposure befalls workers carrying out their duties in a normal way and with the particular equipment assigned to them. The ways of measuring occupational exposure to pesticides are described in Chapter 2 and the ranges of likely exposure to various pesticides are given in Chapter 3.

REGULATORY SCHEMES

In most countries statutory force is given to regulation of the sale, supply and sometimes the use of pesticides as scheduled, or defined. Requirements are set out, often in great detail, for the battery of pre-marketing toxicity testing that must be undertaken, while international agencies such as FAO/WHO and the EEC endeavour to achieve some uniformity by resolving the divergencies that might and do emerge when such action is taken separately by national governmental bodies.

Further, these regulatory authorities, guided so far as possible by experts in their respective fields, then usually promulgate directions relating to each pesticide, stipulating the target pest(s) at which it should be aimed, the crops, foodstuffs or other commodities so to be treated, the amounts to be applied, the manner in which this should be carried out, the precautions to be taken for safeguarding the worker (and the consumer and the environment), time limits to be respected before workers attempt re-entry to the areas treated and any other reservations that may seem necessary. Formal approval may embody categorical instructions covering all these points. These are communicated to users by publications and, especially, by ensuring that they are carried on the labels of the containers in which the material is supplied.

The degree to which these rules are enforced depends very much upon the resolution and the resources of the authority responsible in each sovereign territory, together with the appropriate expertise at their command.

MONITORING AND SURVEILLANCE

In the face of human imperfections, disregard and carelessness, coupled with the inevitable limitations governing the state of the art in relation to toxicological characterisation and hazard assessment, no guarantee can ever be advanced that any proposed use of a pesticide will be totally free from mishap, or danger.

One arrangement that may mitigate the risks is to progress the authorised release over a series of stages - trials clearance, limited clearance, provisional clearance and commercial clearance in the case of Great Britain. These stages are subject to varying degrees of restraint and monitoring. The issuing of approval for full commercial marketing incorporates any conditions added by virtue of the experience gained during the preliminary authorisation.

Even so, indefinite surveillance may still be a wise move. To some extent this can be implemented as an adjunct to enforcement, for when a worker has fallen ill or, worse still, has met his (or her) death because of a pesticide, it is not sufficient just to punish the miscreant. It is essential, so far as possible, to ensure by suitable intervention that the same calamity is prevented from occurring again.

In some of the more dangerous pesticide operations, the workers may need regular medical checking with, wherever practicable, biological monitoring, e.g. blood cholinesterase estimations on those repeatedly handling organophosphorus insecticides.

Finally, all illnesses plausibly attributable to pesticides should be referred to a competent medical authority for investigation and validation - besides ensuring that the correct treatment and management are immediately extended to the patient(s). Only in this way can worthwhile statistics be assembled to arrive at the true incidence of pesticide poisoning. Some health statistics and records of adverse effects of pesticide on health from various developed and developing countries are compared in Chapters 4 and 5.

The enquiries about pesticides to poisons control centres in different countries evidently are not to be equated with the number of actual poisonings. In addition to the accidental non-occupational poisonings and the cases of occupational over-exposure there are, particularly in some countries, a considerable number of suicides involving pesticides. Estimates of the world wide incidence of pesticide poisoning, using so-called statistical models to extrapolate from selected national information, are described in Chapter 5. However, no matter how ingeniously devised, such models may prove positively misleading and the importance of proper epidemiological surveys conducted on defined groups of workers, comparing their pathological vicissitudes with those of carefully matched control populations, is discussed in Chapter 4.

CONCLUSION

Assessing the occupational hazard of a pesticide requires an evaluation of many inter-related pieces of information. This hazard can be summarised, at the risk of undue simplification, as a function of the intrinsic toxicity and the particular method of use. A margin of safety is required by authorities for both the application of pesticides and the resulting residues in food and the environment. The following chapters explain how the margin of occupational safety can be determined from a knowledge of the occupational exposure. Later chapters examine the records of adverse effects on health from pesticides in developed and developing countries and explain the regulatory control of pesticides. The final chapter discusses the adequacy of the practical procedures adopted during pesticide application in public health and agriculture in order to limit personal exposure.

CHAPTER 2

MEASUREMENT OF OCCUPATIONAL EXPOSURE TO PESTICIDES

J.L. Bonsall

INTRODUCTION

The purpose of this chapter is to describe how fundamental information on exposure, and data on toxicity, are employed to control pesticide use and thus ensure the safety of those occupationally exposed in agriculture or public health.

In order to assess the safety of farm-workers or public health workers, their likely exposure to pesticides by skin contamination or inhalation must be evaluated. Many operator exposure studies have been performed over recent years and it is possible to assess safety of one pesticide by comparing it with another, provided that it is applied to the crop or in public health work in a similar manner using the same type of equipment. Details of many such exposure studies are given in Appendix 1.

When a pesticide is to be applied using novel equipment or different conditions which cannot be compared with those in other studies, it may be wise to carry out an operator exposure study. This applies particularly to new means of application and when fundamentally new types of chemical formulation are proposed.

When exposure information is available it is possible to assess worker safety by calculation and examples of such calculations are given later in this chapter.

It is appropriate to define at this stage certain words which are popularly used indiscriminately, but in scientific parlance have clear meanings.

TOXICITY

Toxicity is the innate capacity of a substance to cause damage to living organisms. Toxicity is characterised in terms of acute or chronic effects, and local or systemic effects.

Toxicity by a particular route (e.g. through the skin - effects from dermal uptake, or via the lungs - effects from inhalational uptake) depends upon the pesticide formulation i.e. the product. Pesticides are rarely applied or used in the form in which they are manufactured. Instead they are formulated into products which may contain other pesticides, solvents, wetting agents, stabilisers and carriers, and the presence of these materials often modifies the toxicity of the product. Formulations, because they contain only a percentage of the active ingredient, are generally of lower toxicity than active ingredients, but the presence of a solvent may actually increase the toxicity, for example by facilitating penetration through the skin.

HAZARD AND EXPOSURE

Hazard is a function of toxicity, and degree of exposure. Thus even the most toxic or irritant materials can be safely handled without significant risk of poisoning or harm provided that exposure is correspondingly minimised. Hence,

hazard = toxicity x exposure.

Exposure itself is a function of two other variable quantities, the rate of contamination and time. Actual exposure is the absolute amount of the pesticide either landing on the skin (not clothing) and therefore available for absorption, or present in the air and therefore available for inhalation (and possibly ingestion of larger particles or droplets) in a given time.

Pesticide on the skin is not instantly absorbed, and normally much can be washed off. Because pesticide which is absorbed will be distributed to other organs in the body, and metabolised or excreted, time becomes important. For instance, if a pesticide is completely detoxified and eliminated from the body overnight then each working day any exposure is a new event. However, if the pesticide is not broken down rapidly or if it is stored in the body, like DDT, then over the course of continued exposure the pesticide may accumulate in the body.

Exposure can be minimised in many ways. In developed countries where protective clothing is available and culturally relatively acceptable, there is a last line of defence in that exposure is usually reduced to a minimum by this means. Based on a knowledge of the regional distribution of the pesticide over the body surface from a given method of application, and a knowledge of toxicity and of the composition of a pesticide product (since composition affects the ability to permeate the various materials used to make protective clothing) it is possible to specify the most appropriate type of protective clothing. In the UK, for example, the Pesticide Safety Precautions Scheme (PSPS) defines the type of protective clothing needed for a task. This information is transmitted to the user in the form of the label, securely attached to the container.

Exposure can also be minimised by the adoption of good clean working practices, good personal hygiene, and well maintained equipment. In addition training and experience play a big part in minimising exposure.

It must be emphasised that exposure to a pesticide is not synonymous with uptake into the body (i.e absorption). Exposed skin and outer clothing may be contaminated with quantities of pesticide but bare skin is usually washed although this may depend upon local social customs. Unless the clothing becomes wet or saturated with the pesticide then the degree of protection given by work-clothing is very significant. However, neither work-clothing nor specialised protective clothing provides an absolute barrier to exposure. A chemical which reaches the skin, on the hands, arms or face, or penetrates the fabric of the clothes may be absorbed through the skin and taken into the body. Furthermore, clothing may act as a reservoir for continued exposure, unless laundered. Absorption also depends on the nature of the active ingredient (which determines the intrinsic maximum rate of absorption), the concentration of active ingredient in the formulation, the type of formulation and also climatic factors such as temperature, and particularly humidity.

ABSORPTION FROM CONTAMINATED SKIN

Contamination of the skin by pesticide does not automatically result in the absorption of a significant amount of chemical into the body (Table 2.1). Studies which monitor absorption through the skin need to be interpreted with great caution. For example, the amount of chemical absorbed from experimentally contaminated skin varies very considerably between individuals so similar amounts of skin contamination for workers may result in quite varied levels of absorption and excretion (Maibach, 1976). The location of the skin contamination over various regions of the body often affects the degree of absorption, with greater uptake of chemical into the body from the head or neck than from the forearm (Feldman and Maibach, 1967). An exception is the insecticide carbaryl which appears to be readily absorbed from both the forearm and the face (Maibach et al., 1971). Both the identity and the concentration of each of the

TABLE 2.1 Factors Determining the Amount of Chemical Absorbed from Contaminated Skin

ENVIRONMENTAL	PHYSICOCHEMICAL	PERSONAL
Ambient temperature.	Identity of the pesticide active ingredient.	Total area of the contamination.
		Duration of work periods.
	Presence of organic solvents in the undiluted product on the skin.	Region of the body surface affected by contamination.
Humidity.	Presence of wetting agents in the diluted product on the skin.	Individual susceptibility, including age related changes in the skin.
Volume of contamination landing on the skin.	Concentration of the active ingredient in the contamination.	Work clothing and any garments worn for protection providing protection or reservoirs of contamination.
	Solubility of the active ingredient in water and body lipids.	Personal hygiene including secondary transfer (e.g. from hand to mouth) and the frequency of washing.
	Presence of a vehicle (e.g water) providing intimate contact between the skin and the active ingredient.	Defatting of the skin due to contact with solvents or local abrasions in contaminated areas of skin.

chemicals in the contamination on the skin affect the rates of absorption (Scheuplein and Ross, 1974). Increasing the surface area of contaminated skin leads to greater absorption (Noonan and Webster, 1980). As a general rule, the greatest absorption of pesticide from contaminated skin is likely from a large area of skin heavily contaminated with liquid containing a high concentration of the chemical. Absorption of pesticide from skin contaminated with a dry powder or dust containing a low concentration of the active ingredient is likely to be low.

A proportion of virtually any skin contamination is absorbed if the skin is not washed for a long period (see for example Feldmann and Maibach, 1974; Kolmodin-Hedman et al., 1983 a,b; Akerblom et al., 1983; Libich et al., 1984). The amount of chemical which then appears in the blood is then dependent upon individual variation (Gompertz, 1981).

Some active ingredients do not easily penetrate intact skin. Damaged skin, such as abraded or ulcerated skin, is more permeable, due to the absence of the stratum corneum, which is a protective layer of hard dead cells. Skin which is frequently exposed to organic solvents can be damaged by removal of a natural fatty secretion known as sebum, facilitating the entry of toxicants.

Increased temperature and humidity both increase uptake through the skin, the former by causing blood vessels immediately beneath the skin to carry more blood, so that any pesticide in the skin is more rapidly carried away and around the body, and the latter by hydrating the stratum corneum, which renders it more permeable to materials soluble in water.

COMPARISON OF TOXICITY AND EXPOSURE

The assessment of hazard from the proposed use of a pesticide requires the measurement of toxicity, and the measurement of exposure. The former has been a requirement of regulatory authorities for many years, but it is not usual for exposure data to be required before registration, although exposure studies are being requested more frequently. There is a need for data on worker-exposure in various categories of work involving the use of pesticides in agriculture and public health. These categories are defined by the type of equipment, the nature of the formulation of the pesticide and those work-practices which reflect the agriculture or, in the case of public health work, the buildings and surroundings.

A new pesticide is rarely developed for marketing in one country alone, as development costs are very high. It is usual for a compound to be developed for sale in a number of countries simultaneously. There are protocols for toxicity testing which are broadly acceptable world-wide. A benefit from this degree of standardisation is that if tests are carried out to the same standards and to the same protocol it is often possible to compare their results and thus to compare the toxic properties of different pesticides to assess their relative hazards.

Unfortunately, the assessment of hazard to workers does not only require comparability of pesticide toxicity studies. Information on exposure also has to be considered. There is little comparability in operator exposure studies, as few have been carried out to standard protocols. Such a protocol was developed by the World Health Organisation in 1975 and revised in 1982. Lack of standardisation has led to a number of different techniques for monitoring exposure being developed in response to individual problems, with all the expected difficulties of lack of comparability. This makes it appear more difficult to gauge whether one product or method of application is more or less hazardous than another. However, there is a large body of published information on exposure. Chapter 3 and Appendix 1 simplify the comparison of such published studies, by grouping them by method of application.

The assessment of hazard, if not complicated enough, is further complicated by the fact that exposure studies which are carried out under one set of conditions may be inappropriate for the assessment of hazard under other conditions. It is clear that if a number of pesticides are to be compared for safety purposes, the assessment of exposure must be carried out by comparable methods for each pesticide. Whilst it may be possible to use exposure data from one method of application in one country for safety evaluation in another, it is not necessarily possible to simulate conditions of use from one country in another, as social, cultural, agricultural and climatic conditions may vary. The type of agriculture is also important. The use of pesticides on a crop in one country may be quite different from the use of pesticides on the same crop in another country if the pests are different.

Local factors are also important when considering the use of pesticides for public health or vector control, that is killing insects which carry disease. For instance, if a malaria-carrying mosquito species habitually rests under the roof of a house after feeding, then the under-side of the roof must be sprayed in order to kill the mosquito. If the roof is relatively high, then this increases exposure considerably, as less of the insecticide spray will impinge on the roof, and more fall back onto the sprayman. If the particular mosquito species does not settle under the roof then that part of the building need not be sprayed and therefore worker exposure will be reduced.

The most important variable, however, is the method of application itself and it is this which makes the greatest difference to the degree of exposure, and therefore the hazard. For any method of application it is possible to make a number of other generalisations. The handling of the pesticide concentrate in preparation for spraying gives rise to more exposure than the spraying itself, because its concentration is often 50 to 90 fold greater than the actual spray dilution. Similarly, for outdoor use of pesticides, dermal exposure is generally much greater than respiratory exposure, often by a factor of 100 or more. Indeed, except where a higher respiratory exposure is expected, for instance during insecticidal fogging, or fumigation in glasshouses with a gas such as methyl bromide, the data presented in Chapter 3 indicate that the measurement of respiratory exposure is unhelpful due to its relatively small contribution to the hazard. Measuring respiratory exposure can be technically troublesome, and because of the more advanced technology frequently needed compared with measuring dermal exposure, significant costs can be saved by omitting respiratory exposure measurements. This is obviously important for locally initiated work in the least developed countries where the costs may well need to be minimised.

If the product being used has a particular smell (as some do) it is tempting to think that significant inhalational exposure is occurring, but with normal practices this is not the case. Certain solvents and active ingredients used in pesticide products warn of their presence at very low levels in the air by their smell, which can be quite unpleasant. Nevertheless the actual levels present in air are much too low to cause harmful or even measurable effect.

Analysis of concentrations of chemical in the air often involves measurements close to the limit of analytical detection. This can lead to an over-estimation of extrapolated respiratory exposure levels. Thus, although the limit of analytical determination (the lowest quantifiable level) should be set significantly above the levels of any background instrument "noise" and interference from other chemicals (apparent residues in control samples), it is common analytical practice not to subtract such apparent residues from values found in test samples. Consequently, measurements only just above the limit of determination may include a background contribution. Calculation of the inhaled dose involves a multiplication factor to convert the volume of air sampled for analysis to the volume inhaled per hour by a person. Any background contribution to the measured value is therefore similarly magnified. Over-estimation will also occur during this calculation if minute traces of the chemical have contaminated the air-sampling medium during handling. Thus, it is vital to ensure no contact between contaminated hands or clothing and the air samplers.

STANDARD PROTOCOLS

The idea of the standard worker exposure protocol grew out of the WHO Expert Committee on the Safe Use of Pesticides in 1973. At that time organophosphorus insecticides were becoming widely used in vector control programmes (e.g. for malaria), due to the reduced use of DDT. Since these new pesticides were inherently more toxic, adverse effects on health were seen in the teams of spraymen for the first time and it was realised that exposure needed to be measured to assess the safety of vector control workers. This led to the publication in 1975 of a standard protocol entitled 'Survey of Exposure to Organophosphorus Pesticides in Agriculture'. The protocol embodied some of the necessary techniques for assessing exposure and hazard, and in addition because organophosphorus insecticides inhibit the enzymes collectively known as cholinesterase, details were given of how and when to monitor the activity of this enzyme in blood (WHO, 1975).

According to the WHO 1975 protocol, the quantity of a pesticide landing on the skin was to be measured by means of exposure pads fixed to the clothing on various parts of the body (Figure 2.1). Pads were also fixed beneath clothing on the skin to enable the penetration of the protective clothing to be estimated. By means of a formula to relate each exposure pad to a part of the body, the amount of pesticide falling on the whole body could be estimated. Correction factors could be built in for the amount of the body covered by protective clothing, and for the permeation of the protective clothing by the pesticide. The disadvantage of this method is that it requires extrapolation from the relatively small area of the pads to the relatively large area of the body, and this may well give rise to inaccuracy and therefore unreliable data.

The potential for respiratory exposure could be measured according to this protocol by having workmen wear respirators, the filter cartridges from which could be analysed after the field study was over.

The 1975 WHO protocol also gave details of other data which need to be collected, such as medical data, recording of any adverse signs or symptoms experienced by the spraymen, or other exposed people, the quantity of pesticide used, and any effects on wildlife.

FIGURE 2.1 Indoor Spraying with Insecticide for Mosquito Control

Pads on the clothing quantified exposure to insecticide applied in houses by knapsack sprayer for mosquito control in Indonesia during a 1981 WHO safety and efficacy trial.

This protocol for exposure studies was validated in the Sudan using dimethoate, and a good example of its use can be seen in a later study assessing the safety of the carbamate insecticide FICAM® for vector control purposes described later in this chapter.

The protocol was quite widely used in vector control applications, and in 1978 the WHO Expert Committee on the Safe Use of Pesticides recommended the extension of the survey to other types of pesticide (WHO, 1979).

The disadvantages of the 1975 protocol were its bias towards organophosphorus insecticides, with consequent importance placed on the measurement of cholinesterase, and the reliance on pads to show the overall contamination.

Concurrently a number of pesticide manufacturers were developing their own techniques for assessing pesticide exposure. One of these assessed potential dermal exposure by measuring the contamination by pesticide over the body

FIGURE 2.2 Measurement of Potential Whole body Exposure During Grassland Spraying

Air concentrations of pesticide were assessed by means of filters worn in the breathing zone attached to belt-mounted air pumps. Contamination over each region of the body was measured by analysis of contamination of the garments by 2,4-D. (From Abbott et al., 1984).

surface by using overalls and gloves made of disposable materials, instead of using exposure pads (Figure 2.2). The advantage of this technique was that no extrapolation was required from the relatively small pads to the larger whole body surface area. The benefit of using overalls is immediately apparent. All pesticide landing on the body area is collected and available to be measured. Moreover, the area of such a suit is larger than the area of the body by between 30% and 50% - adding another safety margin. The disadvantage with small (10 cm^2) pads is that they may not detect an area of high contamination, which may land elsewhere on the body, leading to an underestimate of exposure. For instance, exposure of the hands is greatly underestimated or missed altogether.

The WHO protocol was revised and updated in 1981 in the light of progress, and republished in 1982 (see Appendix 2 where it is reproduced) (WHO, 1982).

The revised protocol differs from its predecessor in that it is applicable for use with all types of pesticide, and under most conditions. In addition to the use of patches for assessing dermal exposure, details of the disposable overall method are presented.

The concept of biological monitoring was extended from measuring cholinesterase enzyme activity (for certain insecticides) to the measurement of specific metabolites from a range of other pesticides. However, knowledge of the metabolism and excretion of the compound in man is required before such hazard evaluation is possible.

EXPOSURE MEASUREMENT

Exposure can be measured directly by contamination of pads or suits and air, or monitored indirectly, by measuring metabolites in urine, or occasionally blood, or can be inferred by measuring an effect on an enzyme or physiological process. The classic example of estimating exposure indirectly by means of an enzyme is the measurement of cholinesterase in whole blood or plasma in workers exposed to organophosphorus or carbamate insecticides.

The procedures used to take samples in the field for direct measurement of pesticide exposure are described in Appendix 2. Analytical techniques for measuring pesticides in the samples collected in the field may be found in the published literature or are available from manufacturers. Techniques for monitoring airborne (potentially respirable) exposure are well described in the literature, and may be found in textbooks of occupational hygiene such as those by NIOSH (1973) and Waldron and Harrington (1980).

Indirect monitoring of exposure by analysis of urine for metabolites provides the only reliable evidence of the amount of chemical actually absorbed but it requires prior knowledge of the proportion of chemical which is excreted in the urine and of the analytical methods for the particular metabolites. Moreover, it is not possible to measure urinary metabolites when the overalls-method for measuring potential skin exposure is used, due to the considerable degree of protection given by the overall itself. Careful preparation, such as the measurement of a baseline, also is required for assay of blood cholinesterase enzyme activity as an indirect indicator of exposure to carbamate or organophosphorus insecticides. Some relevant information is given in Appendix 2.

It is worthwhile considering some examples of exposure studies involving direct measurement of exposure. A study of exposure to the insecticide FICAM®, containing the carbamate bendiocarb is a good example of direct and indirect exposure measurements used together in a study (WHO, 1981). It was carried out in Indonesia in 1981 using the 1975 WHO protocol. Exposure pads were worn by spraymen while they treated houses during a malaria control programme. Blood cholinesterase enzyme activity was monitored, and urine collected for metabolites (WHO, 1981). During the actual spraying, the levels of bendiocarb in the breathing zone of the spraymen were also measured.

The dermal exposure part of the study will be described, to compare and contrast it with a later study using the 1982 WHO protocol (Abbott et al., 1984).

These two studies also illustrate the relatively narrow safety margins acceptable for public health work with cholinesterase-inhibiting compounds where the alternative to insecticide use in houses is ill-health from malaria. In contrast, the safety margins for crop use of a pesticide with non-cholinesterase toxicity are larger.

When being monitored in the FICAM® study, each operator wore seven exposure pads (a 10 cm x 10 cm square of plastic-backed absorbent paper) attached to specified positions on the body or clothing as outlined in the 1975 standard protocol and illustrated in Figure 2.1. Adhesive tape was used to stick pad 7 to the skin on the abdomen under the clothing, and pad 1 to the hat. The remaining five pads were attached to the overall using safety pins - pad 2 on back between shoulder blades, pad 3 on chest, pad 4 on left forearm, pad 5 on left thigh, and pad 6 on left shin. The pads remained in position throughout each full day's spraying operation.

As anticipated the supervisors had much lower amounts of bendiocarb on their exposure pads than did the spraymen because they were not actually spraying the pesticide, although they supervised closely.

Table 2.2 summarises the bendiocarb levels on the exposure pads worn by the spraymen. Predictably, the pad attached to the bare skin (number 7) and covered by the overall had much less bendiocarb contamination than those pads pinned to the outer clothing (numbers 1 to 6). In general, the pads worn on the hat and forearm (numbers 1 and 4, respectively) had the highest exposure, as a significant amount of overhead spraying was carried out.

A simple formula based on the WHO 1975 Standard Protocol was used to calculate the actual skin contamination. The formula was adapted to allow for the protection given by the clothing which was an overall and hat, a small gauze face mask and canvas shoes.

The amounts of pesticide on the respective pads (in $\mu g/cm^2$) were multiplied by factors equating to the areas of the particular part of the body (in cm^2). The area

TABLE 2.2 Skin Contamination of Spraymen in Malaria Vector Control

	BENDIOCARB RESIDUE ($\mu g/cm^2$) ON EACH PAD Pad number						
	1	2	3	4	5	6	7
Mean*	10.4	5.5	4.1	7.8	4.3	5.3	0.13
Std. Dev.*	8.1	4.8	3.2	5.3	2.5	3.6	0.08
Range**	1.4 to 65	0.6 to 20	0.4 to 16	1.2 to 62	0.8 to 16	1.4 to 27	0.02 to 0.84

* Calculated after omitting any results falling outside 3 standard deviations of mean.

** Full range shown including outliers.

factors for pads 1 to 4 and pad 6 were for the adjacent skin areas not normally covered by clothing in the regions of those pads. The area factor for pad 7 was the difference between the total body surface area (taken as about 1.8 m^2) and the sum of areas of uncovered skin, representing the total area of protected skin. No attempt was made to alter these areas according to the individual stature of the spraymen. The total actual skin contamination (μg/person/day) was summed as shown in Table 2.3.

Skin contamination from the standard 4.5 hours spraying operation ranged from 6.1 to 157 mg (mean 26 mg) of active ingredient for the sprayman, compared with a mean of 1.5 mg for the supervisors. Only about 6% of the skin contamination was absorbed (Mallyon and Turnbull, 1983). There was sometimes an asymptomatic and temporary inhibition of cholinesterase enzyme activity but the characteristic symptoms of cholinesterase inhibition only occurred occasionally. Overall, in this study there was an adequate safety margin demonstrated for the use of FICAM® in controlling the mosquitos which are the vector for malaria (WHO, 1981).

It is important to note that the results in this skin exposure study represent estimates of the <u>actual</u> skin contamination, not the measured level of <u>potential</u> skin exposure, as each pad represents the likely exposure on bare

<u>TABLE 2.3 Calculation of Total Actual Skin Contamination from Exposure Pads</u>

CONTAMINATION ON PAD ($\mu g/cm^2$)		AREA FACTOR cm^2	BODY REGION
Pad 1	x	837	Exposed skin of the face
+ Pad 2	x	100	Exposed skin of the back of the neck
+ Pad 3	x	149	Exposed skin at the front of the neck
+ Pad 4	x	1208	Exposed skin of the left and right hands
+ Pad 5	x	0	No exposed skin on the thighs
+ Pad 6	x	800	Exposed parts of the feet
+ Pad 7	x	15207	The area of the body protected by garments

skin; for example the pad attached to the hat is used to estimate the skin contamination on the face. However, if hand contamination was greater than the contamination of the pad on the forearm, from which it was estimated, then the total actual skin contamination may have been appreciably underestimated. Trials have been conducted to provide a direct, side-by-side comparison between the exposure pad and overall analysis methods for measuring contamination of skin and clothing (Chester and Ward, 1983). Knapsack spraying caused highly localised contamination which caused the pad technique to overestimate the total exposure. A hand-held ultra low volume sprayer caused relatively more uniform contamination which the pad technique underestimated due to the distribution of the exposure. Chester and Ward (1983) concluded that the overall-analysis method gave accurate determinations of the potential skin exposure.

An example of the use of the 1982 WHO exposure study protocol was a study carried out in the UK with a commonly used herbicide 2,4-dichlorophenoxyacetic acid (2,4-D) (Abbott et al., 1984). No attempt to measure the uptake was made, as this study was designed to measure the total and regional potential dermal and respiratory exposure of agricultural workers during the application of 2,4-D, through five types of application equipment typically used in the United Kingdom. Knapsack sprayers with hand-held lance and single nozzle or hand-held boom with several nozzles were compared with tractor-powered boom sprayers equipped with hydraulic nozzles or controlled droplet applicators. Exposure was regarded as "potential" as the body was covered by the overalls etc. (except the face, for which no measurement can be made by this method) and no measurement of penetration of the overall by the chemical was made.

Potential skin exposure was monitored during the filling and loading operations, and during spraying. Each operator put on a clean disposable overall, socks and gauntlets before the start of each replicate of loading the tank of the sprayer or operating the sprayer. Head exposure was monitored by wearing a hood during spraying.

Filling the tank of the tractor-powered boom sprayers gave rise to potential skin exposure of between 36 and 310 mg active ingredient/person compared with a potential skin exposure of between 8 and 133 mg active ingredient/person while spraying. The replicates were standardised to involve handling the same amount of herbicide when filling each sprayer. Each spraying replicate treated a similar area of land, taking about an hour. Normal work clothing, and gloves or other protective garments used during specific tasks (such as pouring the undiluted product into the sprayer tank) obviously provide significant protection from the potential exposure of the skin to the pesticide. The actual skin contamination can be calculated according to the areas of bare skin, which might typically include the head and neck and possibly the hands and forearms. In this study, if the person doing the spraying had actually worn an open necked work shirt with sleeves rolled up and no gloves or head covering, the skin contamination could have been about 80 mg 2,4-D if the total potential exposure were 100 mg.

The study showed that the hands were the most exposed parts of the body both during mixing/loading operations, and spraying, accounting for between 33% and 86% of total exposure (Figure 2.3). The use of the 1982 WHO protocol in this trial was important, as hand-exposure was not monitored by the 1975 WHO protocol and this important facet of exposure would have been missed.

Many other studies have shown that this potential contamination of the hands is a general rule. To give an example of the distribution of exposure and the consequent absorption of pesticide, consider the following instance. Of the body surface contamination of workers spraying a cereal crop with a fungicide by means of a boom sprayer, most (99%) occurred during addition of the product to the water in the tank of the sprayer, i.e. during mixing and loading. The gloves normally worn during that part of the job received 97% of the total body surface contamination during both filling the tank and subsequent spraying. A working day of eight hours spent repeatedly filling the tank then spraying the crop was calculated to provide a potential respiratory exposure of, at most, 0.5 to 1.4 mg of the pesticide active ingredient. During that time the exposed skin, which was head and arms, would be exposed to 0.5 to 11.0 mg active ingredient. On the basis of urinary excretion of chemical following tank-filling and spraying in normal work clothes with bare arms (Figure 2.4) it was evident that very little (less than 0.3 mg active ingredient) of the fungicide was absorbed from the skin or by inhalation over the same working day (Longland, 1984).

A significant safety margin is involved in the permitted uses of fungicides or herbicides so that exposure of workers is well below a level likely to cause any adverse effects on health. In contrast to the use of pesticides to control malaria there are alternatives to herbicide application such as manual tilling. Unlike the organophosphorus or carbamate insecticides needed in malaria vector control work, products such as 2,4-D are of relatively low toxicity. Acute poisoning due to over-exposure to most herbicides during spraying is highly improbable. The safety of such products during use therefore rests on chronic exposure being far below an hazardous level, for example as described by Hayes

FIGURE 2.3 Distribution of Pesticide contamination over Regions of the Body (From Abbott et al., 1984)

Tank filling of tractor powered sprayers

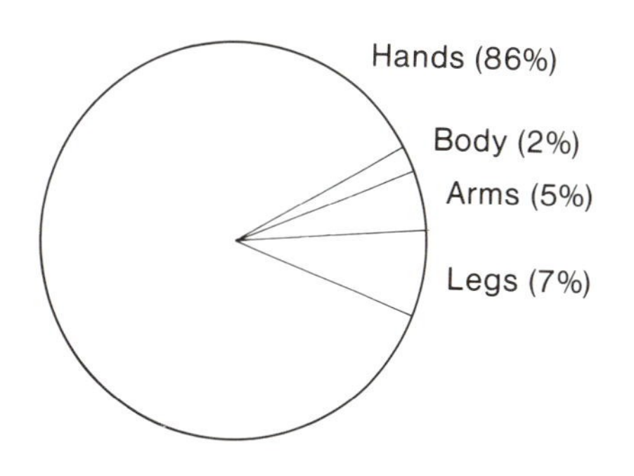

Spraying for weed control using tractor mounted booms equipped with hydraulic nozzles

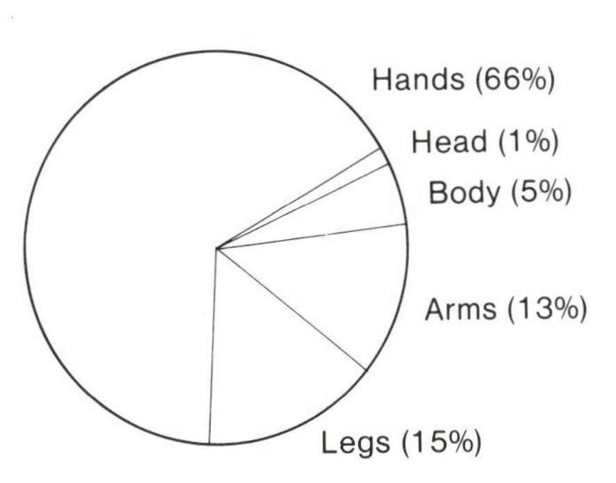

Spraying for weed control using tractor mounted booms equipped with spinning disc controlled droplet applicator (CDA)

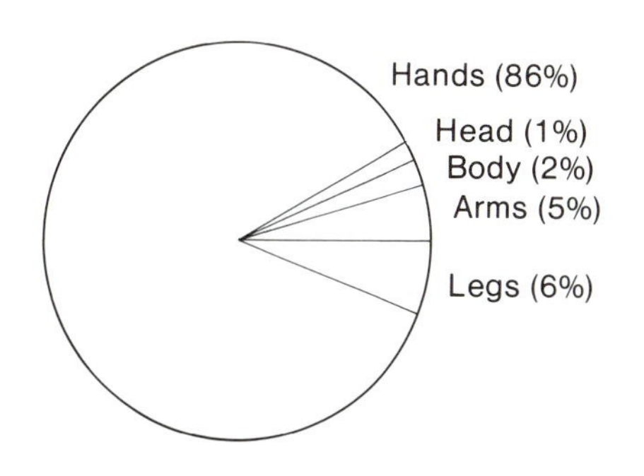

Spraying for forestry weed control using a knapsack fitted with adjustable single nozzle lance

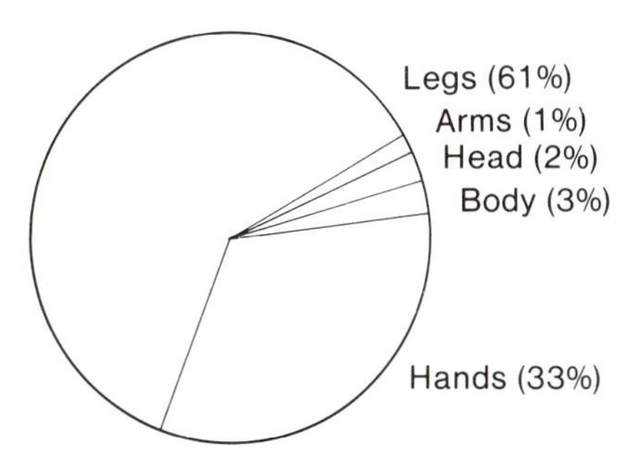

(1982) and Vettorazzi (1979), taking into account important practical considerations and exaggerations deliberately attached to such evaluations (Barnes, 1973).

COMMENT

A primary objective of this book is to bring together as much published (and some unpublished) information on pesticide exposure as possible, in such a way as to allow

FIGURE 2.4 Loading the Tank of a Tractor Trailed Boom Sprayer by Means of a Suction Probe

Contamination of the equipment was evident but repiratory exposure, measured by means of the filter in the breathing zone, with an air pump on the belt, was minimal (Longland, 1984). In addition to the normal work clothes, shown here, a visor or respiratory protection and gloves often are recommended for this task.

informed decisions to be taken on pesticide safety. The difficulty which has been explained is in the interpretation of those studies which have been carried out to very different protocols. It is hoped that those who attempt to study pesticide exposure in future will agree on a consistent method so that studies are more easily interpreted and compared.

This can only have a positive effect on the safety of those who use pesticides, which is of course the reason for carrying out operator exposure studies.

REFERENCES

Abbott, I., et al. (1984) Spray operator safety study. British Agrochemicals Association, London.

Akerblom, M., Kolmodin-Hedman, B. and Höglund, S. (1983) Studies of occupational exposure to phenoxy acid herbicides. In IUPAC Pesticides Chemistry - Human Welfare and the Environment. 227-232. Editor J. Miyamoto et al., Pergamon Press, Oxford.

Barnes, J.M. (1973) Toxicology of agricultural chemicals. Outlook on Agriculture, 7, 97-101.

Chester, G. and Ward, R.J. (1983) An accurate method for measuring potential dermal exposure to pesticides. Human Toxicol., 2, 555-6.

Feldman, R.J. and Maibach, H.I. (1967) Regional variation in percutaneous penetration of ^{14}C-cortisol in man. J. Invest. Dermatol., 48, 181-183.

Feldman, R.J. and Maibach, H.I. (1974) Percutaneous penetration of some pesticides and herbicides in man. Toxicol. Appl. Pharmacol., 28, 126-132.

Gompertz, D. (1981) Assessment of risk by biological monitoring. Br. J. Indust. Med., 38, 198-201.

Hayes, W.J. (1982) Pesticides Studied in Man. Williams and Wilkins, London and Baltimore.

Kolmodin-Hedman, B., Höglund, S. and Akerblom, M. (1983 a, b) Studies in phenoxy acid herbicides.
I. Field Study - occupational exposure to phenoxy acid herbicides (MCPA, dichlorprop, mecoprop and 2,4-D) in agriculture. Arch. Toxicol., 54, 257-265.

II. Oral and dermal uptake and elimination in urine of MCPA in humans. Arch. Toxicol., 54, 267-273.

Libich S., To, J.C., Frank, R. and Sirons, G.J. (1984) Occupational exposure to herbicide applicators to herbicides used along electric power transmission line right-of-way. Am. Ind. Hyg. Assoc. J., 45, 56-62.

Longland, R.C. (1984, unpublished data) Monitoring of spray operator safety during application of prochloraz to cereals in UK. FBC Limited internal report RESID/84/63.

Maibach, H.I. (1976) In vivo percutaneous penetration of corticoids in man and unresolved problems in their efficacy. Dermatologica, 152 (Suppl. 1), 11-25.

Maibach, H.I., Feldman, R.J., Milby, T.H. and Serat, W.F. (1971) Regional variation in percutaneous penetration in man. Arch. Environ. Health., 23, 208.

Mallyon, B. A. and Turnbull, G. J. (1983) Skin penetration and toxic threshold in animals and people exposed to bendiocarb (a carbamate insecticide). Human Toxicol., 2, 552-553.

NIOSH (1973) The industrial environment, its evaluation and control. US Dept. of Health, Education and Welfare, Public Health Service Centre for Disease Control, National Institute for Occupational Safety and Health, Washington.

Noonan, P.K. and Webster, R.C. (1980) Percutaneous absorption of nitroglycerine. J. Pharm. Sci. 69, 365.

Scheuplein, R.J. and Ross, L.W. (1974) Mechanisms of percutaneous absorption, V, percutaneous absorption of solvent deposited solids. J. Invest. Dermatol. 62, 353-60.

Vettorazzi, G. (1979) International Regulatory Aspects of Pesticide Chemicals. CRC press, Florida.

Waldron, H.A. and Harrington, J.M. (1980) Occupational hygiene. Blackwell, Oxford.

WHO (1975) Survey of exposure to organophosphorus pesticides in agriculture. WHO/VBC/75.9. Geneva.

WHO (1979) Safe use of pesticides, 3rd Report of WHO Expert Committee in Vector Biology and Control. Technical Report Series No. 634.

WHO (1981) Safety studies with bendiocarb in a village-scale field trial against mosquitos in Indonesia, 1981. WHO/VBC/81.381. Geneva.

WHO (1982) Field surveys of exposure to pesticides. Standard Protocol, WHO/VBC/82.1. Geneva.

CHAPTER 3

EXPOSURE TO PESTICIDES DURING APPLICATION

G.J. Turnbull, D.M. Sanderson and S.J. Crome

INTRODUCTION

Whenever a pesticide is applied in agriculture or public health there is a recurrence of the situation involving a person, application equipment, safety precautions and the pesticide. These features are common to all work involving the application of pesticides. Differences in terrain, crops, wind speed and the concentration or type of pesticide do not prevent information on worker exposure being grouped according to the type of application equipment and whether the formulation was a liquid, powder or granule.

Furthermore, in any region there is a pattern in the type of farming dictated by the crops, the climate, the degree of mechanisation and the general level of training and education. Likewise the use of pesticides in public health work falls into a number of categories, according to the environment and the pest problem. Hence, studies of occupational exposure to pesticides have similarities which are profound, despite the obvious differences between individual studies.

MONITORING EXPOSURE OF WORKERS

Exposure by inhalation and skin contamination during application of pesticides in agriculture and public health has been measured in many studies. There is a surprisingly large amount of information published in the scientific literature, and additional data reside in the files of regulatory authorities, provided by the manufacturers of pesticides. Recognising the recent trend in several countries to question the degree of contact of farm or public health workers with pesticides, the question really is what facts do we have? The classic approach to any problem is first to review the published literature and group the information; such a review forms the basis of this chapter.

In this review of the published literature, the focus is

on data describing the amount and the distribution of exposure to pesticides that contact the skin or clothing or may be breathed in from the air.

Once the amount of skin contamination and inhalation exposure is known for a particular type of product in a particular use, that information can be employed to predict the likely exposure to other pesticides provided they are being put to similar use. Differences in the active ingredient and the concentration of pesticide employed can be allowed for but only if the method of use is very similar.

In this review, information on exposure is grouped into the work activities that can give contact with undiluted pesticide, followed by the activities involving only contact with diluted pesticide. It may take only a few minutes to load the undiluted pesticide from its container into the sprayer and if necessary dilute it with water. Spraying the diluted pesticide onto the surface or the crop can then take half an hour or an hour before the sprayer has to be refilled. In terms of contact with the chemical one drop or splash of the undiluted pesticide gives more contamination of the skin than many drops of the spray dilution from the nozzles on the sprayer. Although there is no such person as a standard pesticide worker, it is sensible to describe exposure to pesticides for a typical person doing the particular activity in the typical way in order that the information is representative. If the typical way of doing a job means wearing gloves to keep the hands clean, that is the way to measure exposure. Equally, frequent problems such as replacing nozzles that become blocked on the sprayer should be dealt with in the recommended local manner which means that the range of likely events and activities are included in the studies. Unequivocally bad practices and gross abuses are excluded from typical studies although they can be of interest in their own right and must be taken into account when assessing the hazard of the worst likely, rather than the typical or average, circumstances.

To compare exposure resulting from different work activities involving use of pesticides, there has to be some constant time factor. While a farm worker may well spray an orchard or a crop for a complete hour, it is unlikely that one person will spend an hour filling farm sprayers except perhaps in contract spraying operations. However in, for example, insecticide spraying for malaria control, one person may well spend all day handling the undiluted pesticide. It is the number of mixing and loading operations in a work period which is important. For ease of comparison the exposure data in this review have been compiled to describe exposure of one typical person in an hour, whatever the activity. Activities involving handling the undiluted pesticide during loading of the spraying equipment and during the process of diluting the pesticide with water have exposure defined in terms of milligrams of formulation per person per hour of work (mg formulation/person/hour). Contamination by diluted spray is defined in terms of millilitres of spray/person/hour.

The intention in any study of worker exposure to pesticides is to quantify the amount of contact people have (or may have) with the chemical during different activities at work. Contamination of the skin, for example the face, hands or arms, is of greater or at least more immediate importance than contamination of clothing or protective gloves. Even when the skin is contaminated by a splash or by wetting of clothing, normal personal hygiene should ensure that chemical reaching the skin stays there for only a limited time.

POTENTIAL OR ACTUAL EXPOSURE

Even in the hottest weather farm and public health workers handling pesticides wear some clothing and this prevents some of the contamination reaching the skin. Studies on worker exposure to pesticides therefore deal with the potential for skin contamination. Potential skin exposure is the amount of pesticide collected on the exposed skin, and on the clothing, protective gloves etc. that would theoretically reach the skin in the absence (or complete penetration) of garments.

The tables in this chapter show the estimated actual skin contamination, recognising the protection afforded by normal clothing, overalls, gloves etc. The text provides additional information on potential skin exposure which represents an extreme worst case model. From such worst case measurements the amount of actual skin contamination can be calculated for a particular set of circumstances. For example, if only half the potential skin exposure lands on bare skin and the remainder never contacts the skin because of the work clothes and there is no permeation of the clothing, then actual skin contamination is one half the potential skin exposure. Provided there is information on the rate at which the pesticide is applied, the type of clothing worn by the workers, and on the protection afforded by any specially worn protective clothing it is possible to calculate the actual skin contamination from data on potential exposure. The necessity for a simple calculation does not produce a bias. Likewise, from the extensive information on actual skin contamination it is often possible to calculate the actual exposure to other pesticides applied at concentrations different from those in the original investigation. Knowing the distribution of contamination over parts of the body surface associated with the type of task being undertaken allows the actual contamination of the skin under various circumstances to be estimated, for example from the ranges in the tables.

In addition to skin contamination there is the possibility of inhalation exposure. Measurements are made of the concentrations of pesticide in the air in the breathing zone which might be inhaled by the workers. The air sampling methods are best chosen to deal only with the particles or droplets small enough to be breathed in or trapped in the nose or mouth. Again, there is a need to calculate a value for potential inhalation exposure. This is done using information on the measured air levels and a knowledge of the amount of air inhaled by workers doing that sort of work. An alternative method is to measure the pesticide deposited on filters in a mask worn during the period of work. Any contamination of the mask with drops too large to be inhaled gives rise to considerable overestimation of exposure since one splashed drop contains as much as many inhalable drops.

SUMMARY OF WORKER EXPOSURE

Appendix 1 presents data from published literature that describe worker exposure to pesticides, and reports estimates of the amount of pesticide actually or potentially available for absorption through the skin and from the air. The focus is on exposure during application of pesticides in agriculture and public health. There was a paucity of certain information, for example on farm workers or ranch hands putting cattle and sheep through dips or sprays. Information on the possibility for exposure of workers re-entering a crop after it has been sprayed is deliberately omitted since re-entry exposure to pesticides probably is unrelated to the equipment used for spraying. Such exposure may best be controlled by specifying a re-entry interval rather than protective clothing (Popendorf and Leffingwell, 1982). Clothing is of course a most important consideration at the time the pesticide is applied.

The data in Appendix 1 (Tables A to F) provide the definitive details and in this section the intention is to pick out the broad trends and make a general statement on worker exposure to pesticide for each activity. There are startling differences in the amount of exposure according to the work going on. However, the results will be of little surprise to those experienced in agriculture in different parts of the world or who have been involved with public health use of pesticides.

To summarise the data, which in their original form would fill several volumes, the range of values for each of the procedures are provided in the following tables. Naturally there is greatest interest in the typical figures for exposure so the range is for typical information. Extremely high values that fall outside this typical range are rather unrepresentative, even misleading, but they are given (in parenthesis) to show the rare extremes which must be taken into account when evaluating the hazard of the worst likely rather than the typical circumstances. These extreme values represent an unusual exposure which resulted from events specific to the studies involved and it would be quite wrong to compare the procedures on the basis of these unrepresentative values.

Table 3.1 summarises the measured level of exposure to pesticides during loading of the spray equipment and for any mixing with water ready for spraying. The results are given separately for pesticides supplied for use as liquids (emulsifiable concentrates, EC; suspension concentrates, SC; or liquid concentrates), powders and dusts. Different types of equipment are used for dusts or powders and liquids, and the last may be loaded by so-called closed filling systems. In each case skin contamination is much greater than inhalational exposure for either liquid or powdered formulations. Closed filling systems give rise to generally less exposure than open systems. Powder formulations give rise to exposure that generally is higher than arises from handling liquid formulations.

To provide a comparison between the potential skin exposure and the estimated actual skin contamination there is information only from studies involving open filling systems with liquid formulations. The values are 72-2982 (5249) mg formulation/person/h and obviously the potential skin exposure is substantially higher than the estimated actual exposure of 0.70-615 (3258) mg/formulation/person/h shown in Table 3.1.

Worker exposure to diluted pesticides while spraying crops in agriculture is summarised in Table 3.2. The results are organised according to the volume of liquid handled by the equipment and the direction the spray is projected by the nozzles.

Spraying pesticides upwards from nozzles on tractor-mounted, or trailed equipment, as in orchards, (Figure 1.1) gives rise to higher exposure than does spraying with downward pointing nozzles onto low growing crops. As Figure 3.1 shows, a tractor-trailed boom sprayer gives relatively little opportunity for personal contamination during spraying. Respiratory route exposure is almost invariably much lower than the estimated actual skin exposure.

TABLE 3.1 Exposure to Concentrates During Mixing and Loading

	EXPOSURE (mg formulation/person/h)[a]	
	ESTIMATED ACTUAL SKIN CONTAMINATION	RESPIRATORY EXPOSURE
Liquids (EC, SC or liquid concentrate formulations)		
Open filling systems	0.70–615 (3258)	0.004–0.6
Closed filling systems[b]	0.38–17.2*	0.00011–0.03*
Powders	10.6–624 (4960)	<0.22–12.5
Dusts	7300*	15*

a values shown are the typical range from Table A in Appendix 1 with an extreme (worst likely) value in brackets where applicable; abnormal circumstances are omitted.
b closed filling systems may be by can puncture or probe/dipstick type.
* based on very little data.

Pesticide application with air blast equipment may give rise to massive skin contamination. Although there are few data on the potential level of skin exposure it has long been known by experienced workers that the outer clothing and any bare skin is likely to be heavily contaminated during air blast application.

The potential contamination of the body surface during downward spraying is considerably higher than the estimated actual contamination. For example, application volumes between 40 and 1000 L/ha give rise to potential skin exposure of 0.06 to 104 (145.7) ml spray/person/h (from

FIGURE 3.1 Spraying Cereals with a Tractor Trailed Boom Sprayer

The operator applying fungicide to cereals in Eastern England is protected by the cab. Except for loading the sprayer tank and possibly changing blocked nozzles or manually folding the boom there is minimal contact with contaminated surfaces.

Appendix 1) as against an actual skin contamination of 0.003 to 17.3 ml spray/person/h (Table 3.2).

Hand held equipment for outdoor application gives most exposure when large volumes are applied. The estimated actual skin contamination is far less than the potential skin exposure when there are comparable data. For example 50 to 1000 L/ha volume rates give a potential skin exposure of 1.92 to 191.3 ml spray/person/h (from Appendix 1) and an estimated actual skin contamination of 0.016 to 3.1 ml spray/person/h (Table 3.2).

Work in orchards can also involve spraying upwards into the foliage by means of hand-held lances fitted with a nozzle at the end (Figure 3.2). This sort of work gives rise to much higher exposure than spraying downwards with similar equipment onto low-growing field crops. Hand-held spray lances are also used indoors in horticulture and in public health work and some contamination of the skin or

TABLE 3.2 Exposure to Liquid Formulations During Application in Diluted or Undiluted Form

	EXPOSURE (ml spray/person/h)[a]	
	ESTIMATED ACTUAL SKIN CONTAMINATION	RESPIRATORY EXPOSURE
Upwards Directed Application (from Appendix 1 Table B)		
Volume >1000 L/ha (air-blast)	2.2-436 (2021)	0.002-0.4 (2.04)
Volume > 1000 L/ha (not air-blast)	1.37-178.3	0.018-0.12 (0.52)
Volumes 50 - 1000 L/ha	1.02-13.0	0.012-0.039
Volumes <50 L/ha	0.013-1.41	0.0011-0.063
Downwards Directed Application (from Appendix 1 Table C)		
Volume >1000 L/ha	2.6-18.8*	0.002-0.04 (0.65)*
Volume 40 - 1000 L/ha	0.003-17.3	0.00003-0.014
Hand-held Applicators used Outdoors (from Appendix 1 Table D)		
Volume >1000 L/ha	2.0-353	0.002-1.4

clothing cannot be avoided. Work indoors that involves spraying overhead, for example treating the upper walls in houses with insecticide for mosquito control, gives rise to more exposure than work restricted to spraying downwards e.g. onto low crops in a greenhouse.

Exposure to pesticides applied to crops as dusts or granules is summarised in Table 3.3. Results from studies of aerial application are shown separately.

Upward application of dusts predictably gives rise to more exposure of workers than dusting low-growing crops. Mechanical application of granules involves very little exposure but hand application causes greater exposure, which is obvious to any agriculturalist.

TABLE 3.2 (continued)

	EXPOSURE (ml spray/person/h)[a]	
	ESTIMATED ACTUAL SKIN CONTAMINATION	RESPIRATORY EXPOSURE
Volume 50 - 1000 L/ha	0.016-3.1	0.00008-0.012
Volume <50 L/ha	0.096-1.69*	0.002-0.004*
Hand-held Applicators used Indoors (from Appendix 1 Table E)		
Volume 10 -100 ml/m^2	0.19-3.19 (35.1)	0.00007-0.142 (1.91)
Aerial Application (from Appendix 1 Table F)		
Volume 10 - 100 L/ha		
Pilots	0.00088-0.016 (0.096)	0.00005
Groundmarkers	0.11-3.33	0.00006-0.0029
Ground crew, mechanics and supervisors	0.0001-0.042*	No data
Volume <10 L/ha		
Pilots	0.00012*	<0.00001*

a figures shown are a typical range with an extreme (worst likely) value in brackets where applicable. Abnormal circumstances are omitted.

* based on very little data.

During aerial application, dermal and respiratory exposure of pilots is generally low except during dust application. There is no apparent difference in exposure between applications from aeroplanes and helicopters.

FIGURE 3.2 Orchard Spraying with Hand-Held Lance

Spray directed upward into the foliage gives rise to some contamination of the body surface but minimal respiratory route exposure because of coarseness of the spray.

DISTRIBUTION OVER THE BODY

The tables in this chapter and the detailed information on individual studies of worker exposure in Appendix 1 show the work activities which result in skin or clothing contact and inhalation. A number of studies measured the distribution of exposure over the various body regions. Not unexpectedly it was found that handling the concentrated pesticide gives greatest contamination of the hands and forearms and sometimes also legs and feet (see for example Maitlen et al., 1982).

Spraying diluted pesticide upwards contaminates the upper half of the body predominantly, both in agriculture and public health. Downward spraying with a hand-held lance contaminates the lower legs and feet, and with some equipment there is hand contamination, for example when the spray nozzle is being adjusted. Overfilled or leaking knapsack sprayers can contaminate the back.

The fact that workers handling pesticides are contaminated more on some parts of the body than other parts is of tremendous importance. In Chapter 2 the WHO standard

TABLE 3.3 Exposure to Powders, Dusts and Granules During Application

	EXPOSURE (grams formulation/ person/h)[a]	
	ESTIMATED ACTUAL SKIN CONTAMINATION	RESPIRATORY EXPOSURE
Dusts		
Mechanical upward application (from Appendix 1 Table B)	4.3-7.0*	0.058-0.31*
Mechanical downward application (from Appendix 1 Table C)	0.007-1.70*	0.00074-0.0183 (0.041)*
Granules		
Mechanical application (from Appendix 1 Table C)	0.005-0.015	0.000014-0.00046 (0.0018)
Hand application (from Appendix 1 Table D)	0.28-5.88*	0.0004-0.0186*
Aerial Application (from Appendix 1 Table F)		
Pilots - dusts	0.42-5.30	0.001-0.047
Pilots - granules	0.010-0.015*	0.000028
Ground markers - dusts	1.13-2.10	0.003-0.012
Ground markers - granules	0.0434*	0.00042*

a figures are typical ranges with an extreme (worst likely) value in brackets where applicable. Abnormal circumstances are omitted.

* based on very little data.

protocol for exposure studies was described. The need for directly comparable information was explained and some examples were described. Another example is the study carried out in 1983 by the UK Ministry of Agriculture, Fisheries and Food (Operator Protection Group) with spray volumes from 270 down to 54 litres/ha, which is a low application volume (Lloyd and Bell, 1985). Again it was found that exposure of workers operating tractor-trailed sprayers with downward directed nozzles was very small by inhalation. Most of the body surface contamination was on the hands even if there was no handling of nozzles to unblock them during the spraying. Part of the hand contamination resulted from handling surfaces wet with spray while entering and leaving the cab. Other work had shown that the total body surface contamination is higher from operating tractor-powered boom sprayers without cabs than occurs with cabs. However, the contamination was not appreciably more than indicated in Table 3.2. There is some benefit, it seems, from tractor cabs protecting against drifting contamination even when, as is common practice, the cab window is left open to give access to the sprayer controls or for comfort.

INHALATIONAL EXPOSURE

The possibility of inhaling pesticide during work deserves a special comment. Chemical ingested from the lips or inhaled into the lungs is far more likely to be absorbed into the body and reach the major organs than chemical on the skin. However, the evidence is substantial that the levels of pesticide in air provide very little actual exposure (see for example Kolmodin-Hedman et al., 1983 a, b; Akerblom et al., 1983; Libich et al., 1984). Appreciable amounts of pesticide are in the air only when the surroundings are filled with the mist and spray of the pesticide being applied; such conditions occur with mist blowers and the workers involved are well aware of the potential exposure.

To be acceptable for public health use, the pesticide must have little or no odour. Nonetheless as Wright et al. (1981) showed, there are trace amounts of insecticide in the air in treated dwellings. For six different active ingredients the levels in air were between 1.1 and 15.4 $\mu g/m^3$ of air immediately after treatment of the rooms. Over the next three days decreasing amounts remained in the air, down to the limit of analytical detection in some instances. Living in such rooms would provide an inhaled dose of below 0.1 mg active ingredient per day, which is a tiny amount of chemical.

TABLE 3.4 Severity of Actual Exposure to Pesticides During Use - Liquids, Dusts or Granules[a]

(Listed in decreasing order of likely severity)

SPRAYING LIQUIDS	DUST OR GRANULE APPLICATION
Undiluted formulation	
Mixing and loading the sprayers	Loading application equipment with dust
Spray dilution	
Airblast >1000 L/ha	Upward application of dust
Hand-held lance outdoors >1000 L/ha	Hand application of granules
Tractor sprayers directed upwards >1000 L/ha	Pilot during aerial dust application
Boom sprayers directed downward	Ground marker during aerial dust application
Tractor sprayers directed upwards <1000 L/ha	Downward application of dust
Ground marker for aerial spraying	
Hand-held lance spraying indoors or outdoors <1000 L/ha	
Pilot and ground crew for aerial spraying	Pilot, ground crew and ground marker for aerial granule application

a Differences in type and concentration of active ingredient between products and different uses mean that this table is not a rank order of intrinsic safety.

CONCLUSIONS

Activities involved in the application of pesticides give rise to various amounts of actual skin contamination. In Table 3.4 the rank order of severity of exposure is given in terms of the amount of product (as concentrate or dilution as appropriate). The table does not provide a rank order of intrinsic safety. Only a case-by-case comparison of specific products, including their toxicity and intended uses, could rank the margin of safety between actual exposure and the potentially harmful level of exposure. Rather the information in this chapter enables exposure calculations to be made. Attention can be focussed on the activities which produce the greatest actual skin contamination to pesticide active ingredients, and on the distribution of that exposure over the body. The procedures and precautions used to control exposure to a pesticide can be examined in light of this information. Some of the general lessons to be drawn are discussed in Chapter 7.

Use of pesticides in agriculture and public health began long before anyone thought of monitoring occupational exposure; the development of safety precautions and choice of protective clothing was essentially pragmatic. Now, however, crop protection has a scientific basis and, as proposed in Chapter 7, there is a need for a rather more scientific approach to the evaluation and control of occupational exposure during pesticide use.

REFERENCES

Akerblom, M., Kolmodin-Hedman, B. and Höglund, S. (1983) Studies of occupational exposure to phenoxy acid herbicides. In IUPAC Pesticide Chemistry - Human Welfare and the Environment 227-232. Editor J. Miyamoto, et al., Pergamon Press, Oxford.

Kolmodin-Hedman, B., Höglund, S. and Akerblom, M., (1983 a, b) Studies in phenoxy acid herbicides
I. Field study - occupational exposure to phenoxy acid herbicides (MCPA, dichlorprop, mecoprop and 2,4-D) in agriculture. Arch. Toxicol., 54, 257-265.
II. Oral and dermal uptake and elimination in urine of MCPA in humans. Arch. Toxicol., 54, 267-273.

Libich, S., To, J.C., Frank, R and Sirons, G.J. (1984) Occupational exposure to herbicide applicators to herbicides used along electric power transmission line right-of-way. Am. Ind. Hyg. Assoc. J., 45, 56-62.

Lloyd, G.A. and Bell, G.J. (1985) Hydraulic nozzles comparative drift study, Ministry of Agriculture, Fisheries and Food, Operator Protection Group, Harpenden.

Maitlen, J.C. et al (1982) Workers in the agricultural environment: Dermal exposure to carbaryl. In Pesticide Residues and Exposure, 83-103, ACS symposium series, Washington, Editor J.R. Plimmer.

Popendorf, W.J. and Leffingwell, J.T. (1982) Regulating OP pesticide residues for farm worker protection. Residue Reviews, 82, 125-201.

Wright, C.G., Leidy, R.B. and Dupree, H.E. (1981) Insecticides in the ambient air of rooms following their application for control of pests. Bull. Environ. Contam. Toxicol., 26, 548-553.

CHAPTER 4

PESTICIDES AND HEALTH IN DEVELOPED COUNTRIES

J.L. Bonsall

INTRODUCTION

Opportunities for occupational exposure to pesticides can occur during manufacture, transportation or storage and during use in agriculture or public health. Re-entry to treated crops, for inspection or harvest, may also cause exposure. People who are occupationally exposed to pesticides also ingest the small amounts of pesticide residues which may be present in treated crops and are exposed to the pesticides used in public health and in public parks or along pathways. A significant number of people have appreciable occupational exposure to pesticides in agriculture and public health and this chapter considers the effect on their health.

MANUFACTURE

Many countries have pesticide production facilities and so occupational exposure in industry and agriculture can be compared.

Many chemical production plants are owned by American or European multi-national companies, who generally impose high standards of occupational medicine and hygiene, adopting similar standards worldwide. Only rarely are there major occupational health problems. Workers are usually under surveillance by qualified medical and hygiene staff and, when adverse effects occur, they are investigated, documented and frequently are published in the technical press. An important and tragic exception to this good safety record in pesticide manufacture occurred in December 1984 when, according to reports, it appears that over 2000 people died following the emission of the highly irritant toxic gas, methyl isocyanate. This accident occurred in Bhopal, Central India at a chemical plant producing a carbamate insecticide. This gas is used, among other things, as a chemical intermediate in the production of many carbamate pesticides, and is not a pesticide itself. The fatalities mainly occurred among occupants of the huts very close to the chemical plant.

Employers have duties imposed upon them in the form of national health and safety legislation, such as the Occupational Health and Safety Act (1970) in the USA and the Health and Safety at Work Etc. Act (1974) in the UK. Whilst such legislation has national differences, many of the principles are common internationally. Employers in the developed countries are required to take care of the health and safety of employees at work and to have working practices which are safe and do not hazard health or the environment. Information has to be made available to the work force, and frequently the work force is entitled to joint consultations with employers on health and safety matters.

PESTICIDE USE IN AGRICULTURE AND PUBLIC HEALTH

There are considerable similarities between agricultural and public health uses of pesticides in the developed countries and in the developing or least developed countries. Hence, an assessment of the health effects of contact with pesticides during work in agriculture or public health is, to an extent, global although the emphasis in this chapter is on the developed countries and in Chapter 5 on the developing and least developed countries. The developed countries which serve here as examples are in Europe, North America and Australia, and the range of agricultural practices and techniques encompass virtually every type of crop and method of application of pesticides.

A significant difference between agriculture in the developed countries and the developing countries is the frequency of subsistence farming involving small plots of land. In the developed countries, produce is grown for sale, with perhaps only a small proportion of the produce being withheld for consumption by the farmer and his family. Additionally, in the developed countries, farming is commonly highly commercialised, and large plots of land which often are owned by large cooperatives or commercial enterprises and run by a manager and his staff, like many businesses. True there are small farms which are worked entirely by the farmer and his family but these are not at the subsistence level encountered in less developed countries.

Farming in the developed countries tends to be highly mechanised, as machinery is relatively cheap and easily obtained, and labour relatively expensive. Innovation is readily accepted, as farming is profit-orientated, and excesses of production have been, until recently, readily diverted into other markets. New ideas, being rapidly accepted, create an energetic technical press which has in turn fuelled the innovative process. Governments utilise technical advisors to ensure the efficient dissemination of technical advice on how to increase yields. Government departments also commission research from universities or other institutions, again with the object of increasing yields or reducing costs.

Farm-workers are generally well educated, frequently until 16 years old, and many undergo further educational courses at specialised institutions dealing with agriculture. They have a good standard of living, compared with that of farm workers in developing countries.

Thus, in the developed countries there exists an efficient, highly organised and mechanised farming industry, staffed by workers who are generally well trained and usually employed full-time.

All of this is superimposed into countries which have long possessed good standards of internal communications, such as roads, railways and radio, television, telephone and the press. Good medical care is almost universally available. Thus pesticide users are in a position to read and understand labels, obtain protective clothing and, in the event of a mishap, obtain appropriate medical care. Furthermore, the presence of responsible departments in national governments means that mishaps with pesticides can be investigated appropriately and lessons readily learned.

RECORDS ON HEALTH EFFECTS OF PESTICIDES

As such large quantities of pesticides are used, and given the undoubted and undisputed ability of pesticide over-exposure to cause harm, what is the evidence for significant or measurable harm to health by poisoning, including delayed effects such as cancer? A significant number of people in the developed countries have appreciable pesticide exposure, either in manufacture or in use. Given the degree of medical expertise and the sophisticated communications in developed countries, it should be a relatively easy matter to identify appropriate and accurate health data which would be readily accepted as factual by all interested groups. Unhappily this is not the case. Official figures for pesticide poisoning - acute events - are difficult to compare between countries as different data bases are used. National requirements for reporting adverse effects from pesticides vary greatly from country to country. Similarly, there is a dearth of reliable epidemiological studies carried out on farm workers, and only a few on pesticide manufacturers. National mortality statistics for different occupations do not help a great deal as the categorisation of employment groups tends to be wide, and even amongst those categories related to agriculture, not reflective of actual pesticide usage or exposure, despite the undoubted great care taken in collecting and analysing the data.

ACUTE ADVERSE EFFECTS

To give some examples of the scale of adverse effects caused by pesticides, it is appropriate to compare and contrast the experience of some countries including Great Britain and the USA, as much work on pesticides and health as been carried out in these countries. It is not possible to make an extensive international comparison because of the variable standard of national information on poisonings.

Europe

In Great Britain, there is no compulsory official notification scheme for adverse effects from pesticides, except for a limited number of active ingredients specified under the Poisonous Substances in Agriculture Regulations (1984). Such active ingredients are subject to special conditions because of their toxicity, and detailed regulations exist regarding protective measures, such as clothing, hours of work, records etc. Because of the stringent conditions attached to their use, these chemicals tend to comprise only a small fraction of total pesticide usage.

Despite this lack of a formal pesticide adverse effect reporting system in the UK, incidents are reported and are investigated. Poisoning occurrences reported to the Agricultural Safety Inspectorate are investigated by means of a visit from a member of the Inspectorate, accompanied or followed by a visit from a medically qualified member of the Employment Medical Advisory Service (EMAS).

This process occurs whenever a complaint of poisoning by any pesticide, including those not specified in the Poisonous Substances in Agriculture Regulations, is made. Incidents are tabulated and published annually. Since 1981, such officially recorded incidents have been subject to review by another, essentially medical, panel known as the Agrochemical Poisoning Appraisal Panel (APAP). This body consists of members representing EMAS, the official Pesticides Safety Precaution Scheme and the Poisons Unit at Guy's Hospital, London. This extra scrutiny, together with the facility to call for further investigations, gives reliable validation of reports of poisoning.

For Great Britain as a whole, over the years 1978, 1979 and 1980 there were reported respectively 34, 17 and 29 cases of non-fatal occupational and non-occupational accidental poisoning on farms, of which 15, 7 and 19 were attributed to pesticides, 8, 9 and 7 to other chemicals and 11, 1 and 3 to gases. Fatalities numbered 1 in 1978, 2 in 1979 and 1 in 1980, all due to gases released from grain and not pesticides, although for classification purposes all are chemical injuries. By comparison, for the same three years, there were 4606, 4085 and 4307 of all types of non-fatal accidents on farms, with 73, 94 and 78 deaths respectively. Many of the fatalities were caused by self-propelled machines such as tractors, accounting for 25, 26 and 16 deaths respectively, and in the same years bulls and other animals caused 4, 5 and 6 deaths, contrasting with none for spraying of pesticides.

To investigate further the findings of the APAP, a breakdown of their work in 1981 may be of interest - 54 cases were reported, and 25 cases were verified as pesticide-related. There were no deaths, and of the 25 only seven were systemic, mainly due to organophosphorus insecticides. Nine cases were skin effects, mainly of an irritant type, and nine were eye injuries, none with permanent disability (Goulding 1983).

Whilst there is clearly no epidemic of pesticide poisoning Great Britain, it could be suggested that there is an element of under-reporting to the official investigatory procedure. Nonetheless, it is felt that all serious and fatal cases are well documented although some, more minor cases of, for example, skin irritation, might go unreported.

In certain parts of Western Europe the Poison Control Centres, which provide emergency treatment and expert advice, are a major source of information on pesticide poisoning. Examples include Austria, Belgium, France, Great Britain and Sweden. Norway has an official register to which all cases of poisoning are reported by hospitals and doctors and Switzerland collects data from Poison Control Centres, doctors and manufacturers. Employers' medical insurance data provides some information on pesticide poisonings in West Germany.

Information on pesticide poisoning from all these countries has been collected and summarised (GIFAP, 1980). Although there is a need for more recent information a consistent pattern can be seen in Europe. Poison Control Centres receive a great many telephone enquiries about possible poisonings and incidents involving a wide variety of substances. Usually less than 5% of all these enquires involve pesticides. Few of the enquiries that do involve pesticides relate to medically confirmed poisoning with systemic effects or definite local irritation due to skin or eye contact. Fatalities involving pesticides are notably rare and are fewer than other causes of accidental deaths in agriculture.

Two groups of people suffer most from the harmful effects of pesticides. These are children under the age of five years who accidentally ingest the substances due to the carelessness of adults and adults who attempt suicide. Illness due to occupational over-exposure is not common and fatalities due to occupational over-exposure are rare, typically only one or two annually even in large European countries. Whilst no one would regard even this record as completely satisfactory, it must be stated that many of these cases have occurred when label instructions were not obeyed.

Australia

Australia provides more detailed information on poisonings involving pesticides, including the victim's age and the severity of the case (GIFAP, 1980). Pesticide and fertiliser-related poisonings are stated to represent 3% of the total, a very small number being serious or fatal.

ADVERSE EFFECTS FROM PESTICIDES IN AUSTRALIA 1971-1976

	1971	1972	1973	1974	1975	1976
Age 14 and under	321	262	202	321	352	351
Age 15 and over	22	27	20	9	8	14
Age unknown	25	9	13	25	5	34
Total all ages	368	298	235	355	365	399
Deaths	1	1	0	0	1	2
Seriously ill	12	12	1	1	3	3
Slight symptoms/ no symptoms	355	285	234	354	361	394

It was not stated what percentage of deaths were occupational or suicide. From the table above it can be seen that children are most affected.

Canada

Data for 1975 were presented by the Federal Poison Control Centre, which collects data from the Provinces of Canada. Out of a total number of poisonings of about 54000 nationally, 1300 (2.4%) were due to pesticides or fertilisers, 90% occurring in children or young adults. One hundred and fifteen were hospitalised, one fatality being a suicide. It was not stated how many cases of hospitalisation were due to occupational exposure (GIFAP, 1980).

United States

Contrary to what might be expected, the USA has been slow to adopt a centralised system for collecting information on pesticide adverse effects. Account has been taken of fatalities only since 1968 although in 1979 agreement was reached on a pilot programme to report all information received by the Consumer Product Safety Commission (CPSC) to the Environmental Protection Agency (EPA).

In theory, the CPSC receives data from 700 hospitals, including all information and official reports on all cases treated, whether occupational or not, and whatever the outcome. In this arrangement the EPA would be sent full details of each case regardless of severity.

The relative paucity of official statistics in the USA seems paradoxical in view of the activities of the Environmental Protection Agency and the Occupational Safety and Health Administration (OSHA). These government agencies, which devise regulations for the protection of the consumer and farm worker, possess the means to investigate adverse effects and to control the use of pesticides. Given the extent of regulatory activity to control pesticides in the USA, there is a surprising lack of proper epidemiology on a Federal basis.

Studies on adverse effects from pesticides are carried out by various groups, sometimes following an official request. A significant study was reported upon by Hayes and Vaughn (1977). This study examined mortality statistics in selected years over a 20-year period, supplied by the National Centre for Health Statistics. The aim of the study was to examine critically each case reported and to verify the cause of death. According to this study, the number of fatalities (including suicides) attributed to pesticides was:

Year	1956	1961	1969	1973	1974
Fatalities	152	111	87	61	52

and of the fatal cases registered in 1973 and 1974, 35 and 27 respectively were due to malpractice, such as using soft-drink bottles to store pesticides, failure to close containers of pesticides leading to contamination, failure to wear prescribed protective clothing, and entry into grain silos without protective clothing during fumigation.

In 1973 and 1974 only 5 and 7 deaths respectively occurred occupationally in farm workers, again due to not following the official precautions. No deaths occurred when a pesticide was used as directed.

The cause of the decline in fatalities was largely due to a change in pesticide usage, away from arsenicals, inorganic phosphorus and the more toxic organophosphorus insecticides. Additionally and most importantly, it was stated to be due to better awareness on the part of those who applied the pesticides.

In 1976, the EPA published a report on pesticide poisoning cases in which hospital admission had occurred (EPA, 1976). This study was carried out between 1971 and 1973, collecting data from samples of 7000 general hospitals. The results of this study are, however, statistically-derived and are extrapolated from data collected from a relatively small sample of the 7000 hospitals. It was estimated that the total number of hospital admissions due to pesticide 'intoxication' for the three years was 8248, or 0.8% of all hospital admissions, with California having a high percentage of the total. About 1400 cases of adverse effect per annum were estimated to be due to occupational exposure although severity data are not presented (Savage, 1976). About one third of the total number of cases were said to be due to organophosphorus insecticides.

In California, the State Workmans Compensation Law requires that a comprehensive report has to be provided by the physician for every employee treated for a work-related condition. This law, coupled with the relatively large agricultural population in California, means that there are extensive statistics available on adverse effects from

TABLE 4.1 Agriculturally Related Pesticide Illness in California 1973-1978

OCCUPATION	1973	1974	1975	1976	1977	1978
Pilots	14	17	7	8	7	8
Flaggers	20	6	16	14	15	24
Mixers/loaders	165	141	143	122	143	142
Ground applicators	424	225	264	254	236	163
Field work reentry	157	112	167	156	184	95
Gardeners (commercial)	66	103	106	159	155	138
Nursery and greenhouse	112	73	90	119	72	69
Fumigator (field)	71	29	22	14	16	20
Machine cleaning and repair	22	28	40	26	33	21
Tractor drivers, irrigators	-	23	22	30	29	30
Drift exposure (persons nearby)	26	22	31	29	30	44
Total occupational	1077	779	908	931	920	754

After Kilgore and Akesson (1980).

pesticides from that state. From 1973 to 1978 there was a drop in the number of cases reported from 1077 to 754 (Table 4.1).

Table 4.2 shows that 'ground applicators' who apply pesticides using a variety of spraying and dusting equipment have the highest incidence of reported adverse effects. This is being minimised by 'closed systems' for the preparation of sprays containing the more toxic pesticides, and by specifying protective clothing for use with the most toxic pesticides.

Gardeners and nurserymen also have a high incidence of adverse effects, possibly because they are vulnerable, being involved as they are throughout the year with all stages of pesticide exposure, from mixing the sprays and spraying to entering treated crops.

A population which has received much attention are field workers, whose only exposure to pesticides occurs when they enter treated crops. Despite strict controls (see Chapter 6), there is a low incidence of organophosphorus poisoning from surface residues in groups of workers harvesting fruit in California. On average there is one group case reported per annum involving as many as 25 workers and other workers

TABLE 4.2 Illness of Employed Persons Reported by Physicians in California as Due to Exposure to Pesticides in 1973, 1974 and 1975

	SYSTEMIC			SKIN			EYE			EYE AND SKIN			TOTALS			
OCCUPATION	1973	1974	1975	1973	1974	1975	1973	1974	1975	1973	1974	1975	1973	1974	1975	%
Ground applicator	187	80	96	103	66	97	121	66	65	13	13	12	424	225	270	20
Mixer, loader	121	73	74	19	19	20	22	40	34	3	9	3	165	141	131	10
Field worker exposed to pesticide residues	45	12	28	94	78	114	18	12	20	0	10	3	157	112	165	12
Gardener	14	16	24	16	31	28	34	52	41	2	4	14	66	103	107	8
Nursery or greenhouse worker	18	11	25	71	47	54	22	13	19	1	2	2	112	73	100	7
Formulation plant worker	41	55	41	15	4	5	5	10	8	2	2	2	63	71	56	4
Warehouse worker, truck loader	33	25	19	8	10	15	9	10	10	1	3	1	51	48	45	3
Structural pest control worker	11	20	7	5	8	14	8	11	14	0	0	0	24	39	35	2.5
Creosote applicator	1	0	0	24	19	12	9	10	6	2	4	0	36	33	18	1.3
Fumigator of fields	52	7	9	13	9	5	5	12	6	1	1	2	71	29	22	1.6
Cleaner and repairer of pesticide machinery	10	10	10	6	7	9	5	5	12	1	1	4	22	28	35	2.5
Fireman exposure to fire containing pesticides	41	25	37	0	1	0	1	0	0	0	0	0	42	26	37	2.5
Tractor driver or irrigator[a]		6	9		13	10		4	3		0	1		23	23	1.6
Worker exposed to drift from application site	10	11	8	5	4	9	11	6	14	0	1	0	26	22	31	2.3
Aerial applicator (pilot)	10	13	6	0	2	0	3	2	2	1	0	0	14	17	8	.5
Flagger for aircraft	16	3	10	3	2	4	1	0	2	0	1	0	20	6	16	1.2
Mosquito abatement worker[b]		2	0		1	0		4	1		0	0		7	1	-
Indoor worker[b]			50			13			15			1			79	6
Other type of pesticide user	55	67	82	70	46	38	50	39	41	6	2	3	181	154	164	12
Totals	665	436	535	452	367	447	324	296	313	33	58	48	1474	1157	1343	100

a In 1973, cases in these two categories were included under the category, "Other type of pesticide user."
b 1973–1974 reported under "Other type of pesticide user." Agricultural workers, aircraft and ground, account for 52% of all illnesses.

From Kilgore and Akesson (1980).

have suggested that up to 2500 cases per year occur in California alone (Howitt and Moore, 1975; Kahn, 1976).

MEDICAL EXPERIENCE IN DEVELOPED COUNTRIES

The preceding pages highlight one of the principal difficulties when considering adverse health effects from pesticides. Even in the developed world, data on occupational adverse effects and even fatalities are either not collected, or are collected in a manner which confounds analysis. It would surely benefit manufacturers, regulatory authorities and all special interest groups to have better data on this topic, collected in a standard way.

Notwithstanding this, however, it is clear that the number of acute illnesses or deaths from the occupational use of pesticides in the developed world is not very large - typically in single figures in a year even in countries with large populations such as the USA. Deaths occur mainly when common-sense is not applied and labels and leaflets are not read. Systemic poisoning is rare and usually occurs due to misuse of a pesticide. Similarly, other adverse effects such as skin rashes are not common in the agricultural use of pesticides.

CHRONIC ADVERSE EFFECTS

Whilst it is clear that there is no epidemic of pesticide poisoning, there are still fears in the minds of many that pesticides are causing other problems such as cancer or effects on reproduction.

Historically arsenicals were widely used as pesticides. Arsenicals are known to cause lung and skin cancer in man and there is an excess of lung cancer in some workers who have handled arsenical pesticides (Mabuchi et al., 1979). There is now virtually no use of these compounds in agriculture.

The rodenticide ANTU (alpha-naphthyl thiourea) used in the 1940s and 1950s contained beta-naphthylamine, a recognised bladder carcinogen, as an impurity (Davies et al., 1982). This was before the days of routine toxicological testing of pesticides and an excess of bladder cancers has been noted among some of the people who used it. ANTU is still used though not extensively and it now contains much less beta-naphthylamine.

One of the problems encountered when trying to gain evidence about the long-term effects of pesticides is that agricultural populations tend to be widely scattered and, most importantly, have a mixed exposure to many pesticides. Whilst this latter factor confounds epidemiology, it protects the worker, who may not be exposed to any one material for a long time. Epidemiological studies are difficult, even in a long-term study of a group of workers in a closed community, who are exposed to only one hazard - even when the adverse effect is specific and easily recognised.

Only a few groups of workers have been exposed to any single pesticide for a long time and so there have been few epidemiological studies of single-exposure groups. For example, formulators of DDT who were considerably exposed to this pesticide have been followed up in the USA without evidence of any adverse effect (Hayes, 1982).

Additionally, mortality studies have been carried out in manufacturing plants, generally with negative results. Even 2,4,5-T, popularly believed to be a carcinogen (although there is no scientific evidence for this) caused no excess of any disease in production workers (Ott et al., 1980). Epidemiological information on workers in manufacturing and formulation plants which handle pesticides is fragmentary. The prevailing medical opinion is that workers in such plants generally do not have health problems which can be related to pesticide exposure, but further studies are needed. A critical review of the subject would, however, exceed all the space for this chapter.

Another approach is to question whether farm workers have an excess of cancer - or for that matter any other disease - which can be related to pesticides? Taking Great Britain as an example of a country with reasonable national statistics, Table 4.3 would suggest that they do not since the figures for agricultural and non-agricultural workers are very similar. This is borne out by Table 4.4 which indicates that for agricultural workers the proportion of deaths from malignancies is similar to the national population. An extensive international comparison can not be made here because of the variable standard of national and regional information.

CONCLUSIONS AND COMMENTS

It can be seen that despite the extensive usage of pesticides in the developed countries, there is sparse evidence of adverse effects from occupational exposure. Nevertheless there is a feeling that more could be done to monitor the incidence and severity of occupational problems and thus identify measures which could further improve safety.

TABLE 4.3 The Four Main Causes of Mortality in Men and Women Aged 15-74 in England and Wales, 1970-72

CAUSE OF DEATH	POPULATION	PERCENTAGE OF TOTAL DEATHS IN EACH AGE GROUP						
		15-24	25-34	35-44	45-54	55-64	65-74	15-74
Malignant neoplasms	Agricultural	6	22	33	37	31	22	26
	National	12	21	30	33	32	25	28
Diseases of circulatory system	Agricultural	2	16	21	36	46	55	49
	National	6	16	34	43	46	52	47
Diseases of respiratory system	Agricultural	2	4	5	7	10	13	11
	National	6	7	7	8	11	14	12
Accidents, poisoning, violence	Agricultural	74	44	22	9	4	2	5
	National	58	37	15	6	3	2	5
Total percentage	Agricultural	84	85	81	89	91	92	91
	National	82	81	86	90	92	93	92

Calculated from OPCS (1978).

TABLE 4.4 The Main causes of Mortality in Men Aged 15–74; England and Wales, 1970–72 – Agricultural Workers

CAUSE OF DEATH	NUMBER OF DEATHS PER YEAR AT AGES						SMR[a] 15–64	PMR[b] 65–74
	15–24	25–34	35–44	45–54	55–64	65–74		
Malignant neoplasms	5	6	19	71	213	392	104	90
Endocrine, nutritional and metabolic	-	1	-	2	5	12	91	84
Diseases of nervous system etc	1	2	3	3	7	15	115	106
Diseases of circulatory system	2	6	18	86	330	921	85	103
Diseases of respiratory system	2	1	5	21	106	280	122	99
Diseases of digestive system	1	-	3	4	18	39	98	104
Diseases of genito-urinary system	1	1	4	3	8	23	133	116
Accidents, poisonings, violence	61	22	20	25	37	42	163	151
Suicide	6	7	8	8	12	10	190	153
All cases	77	42	77	221	739	1748	103	100

a SMR – Standardised Mortality Ratio – the percentage ratio of the number of deaths observed in the group studied to the number expected from the age-specific death rates for England and Wales.

b PMR – Percentage Mortality Ratio – the percentage ratio of the number of deaths observed from a particular cause in the group studied to the number expected from the age-specific proportions of total deaths attributed to that cause for England and Wales.

From OPCS (1978)

Pesticides are extensively tested for toxic effects and it is difficult to conceive of further safety testing which would improve occupational safety. However, more needs to be done regarding the testing of protective devices, such as gloves, in order to ensure their suitability. Similarly, training in the use and decontamination of protective equipment is needed to ensure its proper use. Indeed, the usefulness of safety training in agriculture cannot be over estimated.

Exposure to pesticides is generally greatest during the mixing/loading operation, most exposure occurring on the hands (Abbott et al., 1984). Pesticide containers should minimise exposure and be compatible with the application equipment - for example it should be possible to avoid having to lift a 20 kg container 2 metres to the top of a sprayer.

Equipment manufacturers need to place more emphasis on design to give clean filling and operation, as well as seeking to fill a niche in the market place with pesticide application equipment in a particular price-band. Containment of pesticide and minimising of operator exposure must not be compromised by market forces.

Finally, better data need to be kept on authenticated pesticide adverse effects, on a national level, compatible with other countries to allow international comparison. Additionally, studies on the long-term health of pesticide manufacturers (who have the highest exposure) and pesticide applicators need to be carried out to confirm that no long-term harm occurs from occupational exposure.

None of the above is beyond the means of any developed, and many developing, countries.

However, such programmes tend to receive low priority, especially in times of economic recession, given the relative absence of any evidence of real harm from occupational exposure to pesticides.

REFERENCES

Abbott, I., et al. (1984) Spray operator safety study. British Agrochemicals Association, London.

Davies, J.M. et al. (1982) Bladder tumours among rodent operatives handling ANTU. British Medical Journal, 285, 927-931.

EPA (1976) National study of hospital-admitted pesticide poisonings. EPA April 1976.

GIFAP (1980) Enquiry into accidents caused by the use of agrochemicals - unpublished document 5th May 1980. No. C3262, GIFAP, Brussels.

Goulding, R. (1983) Poisoning on the farm. J. Soc. Occupat. Med., 33, 60-66.

Hayes, W.J. (1982) Pesticides Studied in Man. Williams and Wilkins, London and Baltimore.

Hayes, W.J. and Vaughn, W.K. (1977) Mortality from pesticides in the USA in 1973 and 1974. Toxicol. Appl. Pharmacol. 42, 235-252.

Howitt, R.E. and Moore, C.V. (1975) Pesticide injury reporting: One study. Economic Social Issues. Univ. Calif. Cooperative Extension, Davis, CA 95616.

Kahn, E. (1976) Pesticide residue hazards to farm workers. In: Proc. Pesticide Residue Hazards to Farm Workers, pp. 175-188. U.S. Dept. of HEW (NIOSH), pub. no. 76-191.

Kilgore, W.W. and Akesson, N.B. (1980) Minimising occupational exposure to pesticides; populations at exposure risk. Residue Reviews, 75, 21-31.

Mabuchi, K., Lilienfeld, A.M. and Snell, L.M. (1979) Lung cancer among pesticide workers exposed to inorganic arsenicals. Arch. Environ. Health 34, 312-319.

OPCS - Occupational Mortality - decennial supplement (1978) Office of Population Censuses and Surveys. HMSO (London)

Ott, M.G. et al. (1980) A mortality analysis of employees engaged in the manufacture of 2,4,5-trichlorophenoxyacetic acid. Occupat. Med. 22, 47-50.

Savage, E.P., Keefe, T. and Johnson, G. (1976) The pesticides poisoning rate is low. Agrichemical Age 19, 15

CHAPTER 5

PESTICIDE EXPOSURE AND HEALTH IN DEVELOPING COUNTRIES

J.F. Copplestone

INTRODUCTION

In many spheres of activity in the developing world, there is a dearth of hard facts on which to base descriptions or judgements, and the causes and consequences of occupational pesticide exposure are no exception. Administrative structures tend to be based on former colonial patterns, and these were not oriented towards the collection of data that were not of strict (and often commercial) interest to the parent country. In many developing countries, there is no statistical tradition or baseline, and the resources do not exist to establish these. There are more important needs to be met.

Therefore, generalisations on pesticide exposure can only be true in some places and false in others due to the profusion of social, religious, economic and political patterns that exist. However, these different modes of life do mean that there are definable differences in potential pesticide exposures, and these are often little understood in the developed world.

DEVELOPING COUNTRIES

It is generally accepted that about twenty per cent of all pesticides are sold in developing countries. Most active ingredients are manufactured in developed countries and are shipped either in this form for local formulation or as formulated concentrates. The ratio of insecticides, fungicides and rodenticides to total pesticides is higher in developing countries since these are used for the control of the major pests of tropical areas. Of the pesticides destined for non-agricultural uses, many are bought by governments for public health purposes. These are frequently destined for the control of vector-borne diseases of humans which are the scourge of the developing world; most are insecticides for the control of malaria, filariasis, and trypanosomiasis, with some molluscicides for schistosomiasis control. Rodenticides not only protect food stores but also contribute to the prevention of plague. The use of pesticides in public health gives rise to exposure potential both of applicators and the general public.

There are three main types of agricultural activity: plantation farming, which tends to be monocultural and requires intensive pest control; cash cropping which is diverse in both the types of crops grown and the size of the holdings; and subsistence farming. The smaller the holding and the more that the crop is for subsistence, the lower is the likelihood that pesticides will be used. The crop itself also has an influence, the two crops most vulnerable to insect attack being cotton and rice. Herbicides tend to be less used in all types of agriculture, since weed control can be achieved by human effort rather than chemical means.

The level of mechanisation is low and many pesticides are applied by simple means, except in plantations. Pesticide exposure is also influenced by the terrain. In huge river basins, the land virtually floats, and it is impossible to handle pesticides without some localised (but usually insignificant) contamination of the ground water; in other vast areas, there is little provision of water for ordinary human hygiene.

Social patterns influence pesticide exposure in ways not seen in developed countries. In public health campaigns, and in nuisance pest control in municipalities, spraymen are usually employed. The pesticides that they use tend to be limited both as to type and toxicity; the inevitable exposure of the general public in the spraying of households in villages and the use of space sprays and fogs in urban areas limits the choice of pesticides and formulations. On the other hand, the spraymen may apply a pesticide for weeks or months on an almost daily basis. Although they may be employed for long periods, they are usually daily labourers since, in most countries, being a sprayman does not infer any social prestige. Their wages are low and they are more likely than not to be illiterate. They receive training in spraying techniques but safety practices are often neglected so as not to alarm them unduly to the extent that they might seek remedy in additional recompense. Frequently, they receive neither guidance nor good example from their supervisors, but they are usually protected by the low toxicity of compounds recommended for public health use. Only when for some reason the unexpected happens, as in Pakistan in 1976 (WHO, 1978) when some batches of malathion developed unforeseen toxicity, does the lack of good safety practices and protection take its toll.

In plantations, employees tend to be housed, fed, and medically supervised by their employers, and to this extent, their life style tends to approach that of similar employees in developed countries. Plantations vary in the care and training that they provide and sometimes set their own standards rather than conform to any minimum national standard of working conditions. Pesticide usage may be high but the amount of exposure usually depends on the quality of management.

In most cash cropping operations, the situation is different in many respects. The group of people involved is very diffuse and numerous. Many cash crop farms are family affairs. Others employ families to carry out the daily routine, rather than individual workers. Thus, large sections of rural populations are involved occupationally. In some developing countries, these people may make up more than half of the total population, having excluded urban populations, plantation workers and those too old or too young to work in the fields or shepherd the flocks.

Among these large numbers, the standard of education varies over its whole range. The level of educational attainment of individual farmers does not have any correlation with the size of land holdings. Tribal hierarchy is more important. A large proportion of those who make daily decisions on agricultural operations will be illiterate or nearly so. It is sometimes overlooked that illiteracy in rural populations in many parts of the world also implies difficulty in interpretation of pictures. The people have not been exposed to television, which in many developing countries is found only in the cities. Formal education may be limited to religious subjects, particularly in Moslem areas, and this involves learning by rote without any pictorial representations to assist. Real communication exists only through the spoken word, but this does not always depend on personal contact. Radio and telephones are understood and used.

Possibilities of indirect pesticide exposure depend on family structures and traditional living habits. In some countries, farmers and their families live close to their fields, and the risk of indirect exposure is higher than in those countries where they live in villages and walk daily to their work. People in the developing countries may be exposed periodically to high residues on food since, contrary to safety instructions, harvesting shortly after spraying is by no means a rarity. From the monitoring of residues so far reported (FAO/WHO, 1982) there does not appear to be a problem, except perhaps as a rare contribution to cumulation of the effect of pesticide exposure when combined with an occupational exposure to a pesticide of a similar type.

In many countries, the access of rural populations to medical facilities is poor. Hospitals with limited facilities are usually situated in small towns and serve large populations, as do the few doctors in rural areas. In recent years, under the influence of the policy of the member states of WHO "Health for all by the year 2000", there has been increasing emphasis on the provision of basic health workers at the village level. This may do much to raise the level of health of villagers, particularly through immunisation and infant care. It may also serve to alleviate some of the serious and sometimes unsurmountable problems that arise in transporting a patient to medical care from a rural area.

For subsistence farmers, the problem of pesticide exposure is simpler, because of their inability to afford the products. They remain the victims of natural pest disasters and crop losses, except insofar as they may be beneficially affected by pesticides applied in the vicinity. Probably this is the group least exposed to pesticides in the world, but their usual state of poverty and malnutrition does not make their condition enviable.

PESTICIDE TRANSPORT AND STORAGE

The pesticides available in developing countries are of varying quality, and this may give rise to a different pattern of exposure than that which occurs with the same products in developed countries. When formulations are shipped as such, they may be subjected to a variety of abuses en route. The containers may be exposed to prolonged high temperatures in ships' holds, on docksides, airport tarmacs, in railway yards or in the depots of the wholesalers at the end of the distribution chain. They may be soaked repeatedly by torrential rain which rusts metal containers, softens cardboard, and sometimes soaks off labels or makes them almost illegible.

Many plastic bags degrade on storage, becoming brittle and tearing easily. All this results in spillage which directly exposes those who have to handle the pesticide from this point, but also can have catastrophic consequences if food is contaminated during transportation or storage.

More subtle changes can take place within the formulation itself. Sometimes this can take the form of chemical changes and possible enhanced toxicity, but more often it results in degradation of the active ingredient so that the dosage in final use may be far below the intended effective level. In one sense, this reduces exposure since most degradation products are less toxic than the active ingredient, but the physical degradation of the formulation tends to reduce its handling qualities; it may require additional mixing and filtering, and may cause blockages in spraying equipment, thus increasing the hazard of application unless additional precautions are taken.

Formulations made in the country of use may compare with those made in developed countries, but they may also contain inferior materials. Errors may be made in the content of active ingredient, containers may be unsuitable, and labelling may be inadequate or misleading.

PESTICIDE REGULATIONS

Pesticide registration exists in most developing countries but this is usually a bureaucratic exercise, and is not followed by any type of monitoring of compound quality, simply because the analytical facilities are not adequate for this purpose. It is sometimes forgotten that pesticide registration has no reason for existence at all, except for the protection of man and his environment. Systems of great complexity have been devised and refined in the developed countries and have been taken as models in the developing world, without any serious attempt to adapt them to national circumstances. It is not surprising that they are largely ineffective.

Of recent years, the need for harmonisation of registration requirements has been much discussed (FAO, 1982) and a measure of success has been achieved. What is missing from the international discussions is an epidemiological approach to try to ascertain the effects on men of pesticide exposure, and the elements of pesticide registration which have been responsible for their diminution or control. It is arguable that in no other scientific subject has there been so much speculation and extrapolation.

MEDICAL EXPERIENCE IN DEVELOPING COUNTRIES

Despite the amount of laboratory research put into the definition of possible long-term effects of pesticide exposure, it has become clear in epidemiological studies that the major problems are undoubtedly acute exposure effects. Forty years of exposure to organochlorine insecticides, including heavy exposure to DDT and other compounds for the first twenty years, and thirty years of exposure to organophosphate compounds have not yet resulted in the demonstration of any well-documented and unequivocal long-term effects in any highly exposed population.

With a few exceptions, such as alkyl mercurials, acute

poisoning by pesticides results either in death or recovery after a relatively short period. For several classes of pesticides, such as organophosphates, carbamates and anticoagulants, there exist specific antidotes - a situation less common in toxicology than the public usually imagines. Thus, if misuse results in high exposure, the results should be mitigated if medical attention is prompt and the antidotes are available. This is true in the developed countries, but in the developing regions, the outcome of a case is often determined more by absorbed dosage than by the effects of treatment. For this reason, although the quantities of pesticides used in developing countries are proportionately less than in developed countries, there is probably a relatively higher incidence of cases of acute poisoning, and the fatality rate of cases is higher.

STATISTICAL MODELLING

This was taken into account in a statistical model designed to measure the global incidence of accidental (as opposed to intentional) poisoning in the early 1970s. The model utilised the few statistics of incidence available at that time. The figure that emerged - in the order of half a million cases per annum (WHO, 1973) - has since been quoted, doubted, and misused on a number of occasions but has not been disproved. The defects of the model arose, not from any bias, but from an insufficiently broad base of national figures. This situation has not improved much since 1973, but such figures as have become available have only served to confirm the validity and, perhaps, the conservatism of the model (Copplestone, 1977).

RECORDED ADVERSE EFFECTS

One study in particular has indicated the possible pattern of the results of exposure in a developing country. A six-year survey of hospital admissions in Sri Lanka for pesticide poisoning has recently been published (Jeyaratnam et al., 1982). Sri Lanka is a country with moderate to high pesticide usage in both agriculture and public health. Perhaps the only way in which it has differed from other countries is a restriction since 1978 on the use of malathion in agriculture in an attempt to delay resistance to this compound in mosquito vectors of malaria. The mean number of hospital admissions per year for pesticide poisoning over the period averaged 13,000 and in 1979, the year of fewest admissions, the case rate was 79 per 100,000 inhabitants. The case fatality rate was 9.4% overall but varied with the hospital. Seventy-three per cent of all admissions were suicidal attempts (showing a predilection for pesticides which Sri Lanka shares with a number of other countries in the Eastern hemisphere), and the remaining cases were accidental (24.8%), or the cause was unknown. Occupational cases were 17.1% of total cases and 69% of the accidental cases. Males outnumbered females in all categories, but the proportion of females was higher among the suicidal attempts. Organophosphorus compounds figured largely among the agents responsible (76.0%), but no particular pesticide was identified in a rather substantial number of cases (15.9%). The case fatality rate of organophosphorus compounds was 21.8% but no distinction was drawn between suicidal and accidental cases.

The figures quoted above contribute to the global picture, which is now beginning to emerge. As a generalisation, and to give a sense of proportion to the problem, it seems that accidental poisoning by pesticides accounts for about 4-5% of all accidental poisonings. The preventive aspects of accidental poisonings have not, perhaps, had all the attention they deserve, considering the amount of morbidity and mortality caused. Drugs and household poisons figure much higher than pesticides in the list of causative agents. For pesticide poisonings, occupational exposures usually account for 60-70% of all accidental poisonings. The chief victims among the non-occupational cases are children under five years old. The mortality rate of the occupational cases tends to be lower than the non-occupational cases, and the incidence of cases is considerably higher in developing countries.

CONCLUSIONS AND COMMENTS

While the developed countries have been able to reduce considerably the number of cases of pesticide poisoning in the last two decades, it is clear that the incidence in the developing countries is unacceptable. In every case of occupational over-exposure, the root cause is misuse of the pesticide, and this is usually due to ignorance, carelessness, misinformation, the unavailability of adequate protective equipment or the malfunction of application equipment. Moreover, the special circumstances of exposure in developing countries, as outlined above, tend to lower the threshold beyond which misuse may result in overt adverse effects. Age, nutritional status, concomitant disease and parasitic body load may all be responsible for an increased susceptibility to the effect of pesticide exposure.

The classic approaches to prevention in the developed countries have been legislation and education. Probably, in the long term, these will succeed in developing countries also but at present the infrastructure in regulatory and advisory services rarely exists effectively.

Sovereignty of Governments

One approach sometimes suggested is the refusal by developed countries to supply pesticides of high hazard or toxicity to developing countries, except for uses approved in the exporting country. Although superficially this may seem to solve the problem, there are several objections of considerable weight. Firstly, there is the principle of sovereignty which, although sometimes overlooked in politics today, allows an importing country to make its own decisions on desirability of imports. Secondly, even if export restrictions were imposed on a proprietary product, these would be enforceable only as long as the patent lasted or as licences to manufacturers were granted. There are already many commodity (out of patent) pesticides. These may be made in several countries, some of which might not impose similar restrictions: therefore, the ban would be ineffective. Thirdly, a ban may retard the development of formulation plants in developing countries by depriving them of technical pesticides as raw materials. These formulation plants sometimes permit considerable savings in scarce hard currency and, if properly controlled, are to be encouraged in developing countries.

However, the most powerful argument of all is that the conditions for which pesticides may be needed differ markedly between the exporting countries and the developing countries. In the latter, the crops tend to differ, so do their pests. Vector-borne and parasitic diseases of men and animals are very widespread and are only controlled with difficulty. Thus, the approved uses of a pesticide in a developed country may have no equivalent in a tropical developing country. As in developed countries, the use of pesticides on tropical pests has a profound economic impact, as also does the control of parasitic diseases, but it is a sad fact that resistance in the pests has led to the need for different, and sometimes more toxic pesticides. The importing country is, therefore, faced with a risk-benefit equation which might have a very different solution from that in a developed country. Such factors as the effects on nutrition, or on the infant mortality rate may have to be included in the assessment of benefit. Thus, any attempt on the part of a developed country to make decisions affecting a tropical developing country may well have a deleterious effect on the health of its people and on its economy.

Training and Labelling

The fact that a developing country may have a real need for hazardous compounds in no way detracts from its responsibility to see that these are used safely and are not misused in such a manner as to result in cases of accidental poisoning, especially when used occupationally. Possibly the most effective single measure would be the restriction of the most hazardous compounds to teams, governmental or private, who have been specially trained in application and whose work practices and precautions can be regularly checked. Such a restriction could, for example, be applied to all formulations in class 1 A or 1 B of the WHO Recommended Classification of Pesticides by Hazards (WHO, 1984). This does not mean that the restriction should apply to all technical products listed in Class 1 A and 1 B of the Classification. All classification must be by formulation, and formulations of such technical products with low concentration of active ingredients may well fall into lower hazard classes.

Ignorance as a cause of misuse can only be abolished by education. There are many ways in which this can be done: by educating health and agricultural workers in contact with the communities, by radio talks, through farmers' groups or co-operatives, through retailers, or through community leaders. In addition to these, the most practicable source of information, at present, is the label. However, the label required by registration authorities now contains so much detail that it is almost entirely self-defeating. Possibly, there is a need for a separate simple safety label. This should clearly indicate by colour for hazard class and by symbol the precautions to be taken. With the few possibilities at our disposal clearly to distinguish labels (and, therefore, products) one from another, it is inexcusable to use colours to denote use categories or types of pesticide. Red should be used for formulations in Hazard Classes 1 A and 1 B, orange for Class 2, and blue for Class 3.

Similarly with wording, the skull and crossbones symbol should be reserved for formulations in Classes 1 A and 1 B with the words "Very toxic" and "Toxic", as appropriate. The St. Andrew Cross should be used for Hazard Class 2 with the word "Harmful". Class 3 is denoted by the word

"Caution". This simple symbol and wording system has been recommended by FAO and WHO (FAO, 1985). Although pictograms may not be universally useful, they can be added to demonstrate the degree of protection required.

In the absence of national infrastructures for pesticide control, more responsibility for education rests on the vendor of pesticides. It is important that personnel all the way down the distribution chain should be aware of the hazards as well as the uses of the products they are selling. They should pass on advice tempered to the hazard, and should not pretend that hazardous pesticides can be used without taking precautions. The higher the hazard class of the pesticide, the greater is the need of the user for adequate advice and the responsibility of the vendor to give it.

It may be that this could be one of the most effective means of education in developing countries, but a prerequisite is that all pesticide packages, besides being adequate in resisting the treatment they are likely to receive, should be marked with the hazard class.

It will never be possible to prevent altogether cases of occupational pesticide poisoning which are only an expression of excessive exposure. There are always in any society those who will not wear their safety belts in their cars or their safety glasses at work. However, with discipline and education, the number of cases of pesticide poisoning in developing countries can be reduced, and it is incumbent on governments and on pesticide manufacturers and distributors to ensure that effective steps are taken to achieve this.

REFERENCES

Copplestone, J.F. (1977) A Global View of Pesticide Safety, In Pesticide Management and Insecticide Resistance pp 147-155, Editors D.L. Watson and A.W.A. Brown, Academic Press, New York.

FAO (1982) Report of the Second Government Consultation on International Harmonization of Pesticide Registration Requirements, document AGP: 1982/M/5, Food and Agriculture Organisation of the United Nations, Rome.

FAO/WHO (1982) Global Environmental Monitoring System: Joint FAO/WHO Food and Animal Feed Contamination Monitoring Programme - Summary and Assessment of Data Received from FAO/WHO Collaborating Centres for Food Contamination Monitoring.

FAO (1985) Guidelines of Good Labelling Practice, (in preparation), Food and Agriculture Organisation of the United Nations, Rome.

Jeyaratnam, J. et al. (1982) Survey of Pesticide Poisoning in Sri Lanka, Bulletin of the World Health Organisation, 60 (4): 615-619.

WHO (1973) Safe use of Pesticides, Technical Report Series No. 513, 42, World Health Organisation, Geneva.

WHO (1978) Chemistry and Specifications of Pesticides, Technical Report Series No. 620, 7-10, World Health Organisation, Geneva.

WHO (1984) The WHO Recommended Classification of Pesticides by Hazard and Guidelines to Classification 1984-5, unpublished document VBC/84.2, World Health Organisation, Geneva.

CHAPTER 6

PESTICIDE REGULATIONS

R.C. Tincknell

INTRODUCTION

Everyone involved in the manufacture, distribution and use of pesticides has a part to play in the avoidance of accidents. Much thought has been given to devising procedures for handling pesticides so as to minimise risks to health and many publications are available on the subject (GIFAP, 1983).

The observance of safe use precautions, especially for products of high toxicity, is so important that in many countries it is enforced by legal and regulatory provisions, aimed at the protection of those who are occupationally exposed.

The strength of such provisions varies greatly from country to country. In some there are virtually none or merely guidelines, whilst in others there are very specific statutes, often drafted within the framework of more general legislation covering the protection of workers.

This chapter deals with the principles that underlie legislative and regulatory provisions and discusses the ways in which different authorities have put them into effect. It is, of course, important to recognise that such provisions are in a state of active evolution and although this chapter cites current regulations by way of illustration it is not intended to be used as a comprehensive source of regulatory detail. This should always be obtained by consultation of the published regulations and, in cases of doubt, in discussions with the authorities themselves. A most useful, but not exhaustive list of national authorities is presented by the Council of Europe (1984).

It is almost a truism to say that no pesticide would harm anyone if there were no exposure. Similarly there are few pesticides that will not cause injury if exposure is excessive. Consequently, one cannot really speak of a pesticide as being safe or dangerous; one can only speak of safe or dangerous use practices. There are no harmless substances, only harmless ways of using substances (Jeyaratnam, 1982). Harmless ways of using pesticides depend primarily on the avoidance of harmful exposure and safe use precautions are mainly concerned with such avoidance.

In deciding whether a product can safely be used, authorities have to take into account three main issues. Firstly, what is the consequence of failing to keep exposures down to reasonable levels? In some cases there may be wide latitude whereas in others quite small accidental exposures could be dangerous.

Secondly, to what extent can users reasonably be expected to respect the precautions required? Are they so tedious, uncomfortable, or even impracticable as to be unlikely to be put into effect?

Thirdly, in the projected circumstances of use, what dangers are there of excessive exposures arising purely from accidents? To what extent are the intended users likely, for example, to be exposed to failures of equipment, the carelessness of others, or even to their own carelessness?

Thus, in addition to the products intrinsic toxicological characteristics, much depends on the circumstances in which it is to be used. It is perhaps ironic that countries where use conditions are the most difficult are often those with the weakest means of ensuring safe use.

How then do regulations contribute to promotion of the safe use of pesticides? In their broadest sense, pesticide regulations have to cover all aspects of use, including not only operator safety as such, but those other major issues relating to residues in foods, environmental effects, and product efficacy. That aspect of regulations which is the concern of this chapter, the protection of operators, contributes by providing a means to focus attention on the main issues. These are outlined in the following sections.

APPRAISAL OF HAZARD: DATA REQUIREMENTS

The primary data requirements for the appraisal of pesticide toxicity are those concerned with the acute effects of the product on animals by various pertinent routes of exposure. In most cases the data are derived from studies on the rat and are concerned mainly with oral, dermal and sometimes inhalational exposures, although in rare cases, notably Japan (MAFF, 1983) intraperitoneal route data are also called for. In most countries, supplementary data covering such issues as skin and eye irritancy and skin sensitizing effects are also needed. There is a very useful detailed list of the data needed by the majority of countries in the Council of Europe (1984) publication. At the national level, guidelines from Canada (1978), Japan

(MAFF, 1983), the United Kingdom (PSPS, 1983) and the United States (EPA, 1982a) are among the more prominent. Although in most of the developed countries the requirement to appraise the product is statutory, adherence to guidelines for data needs is not always so rigid. However, even where there is considerable latitude in the strict regulatory sense, an authority is unlikely to be satisfied unless its own guidelines have been reasonably well followed so that guidelines do, in fact, carry considerable weight.

Most of the guidelines referred to require the acute data to be provided for both the active ingredient and the formulation to be sold, or at least a formulation reasonably similar to it.

Of course, during the development of a pesticide, many other data have to be developed for the appraisal of such issues as long-term exposure to the low levels that may occur as residues in foods. These data, although they may not be considered as being needed primarily for the purpose of appraising hazard to the operator, are nevertheless taken into account by most regulatory authorities in their appraisals. Most would be seriously concerned at the possibility of operators being subject to low-level but prolonged exposures to products that might give rise to a carcinogenic or teratogenic hazard, for example. Moreover, information relating to the mechanism of toxicity will also be taken into account.

It should, of course, be recognised that practically all of these ancillary data are developed on the active ingredient rather than the formulation. This, in the past, has given rise to a certain degree of confusion in that some authorities have considered that data should also be developed on the formulation to be sold. This, of course, is quite unnecessary, because formulation has relatively little effect on the intrinsic long-term effects of the active ingredient, as distinct from the often quite important effects on the acute toxicology.

As well as toxicity data, some guidelines also mention the need for data on the rate at which pesticide crop deposits decline so that an estimate can be made as to when it would be safe for workers to enter the treated area to perform such tasks as harvesting. The basis of such estimates is the extent to which the worker might accumulate a toxic dose of product by transfer of dislodgeable deposits from the crop onto his exposed skin. This point is made in the guidelines for Great Britain (PSPS, 1983). Popendorf and Leffingwell (1982) reviewed the history of the re-entry question in the USA in regard to organophosphorus insecticides. Again, in the USA, the EPA (1982 b), lists the data that may be needed to decide when it would be safe for workers to enter treated fields, but the data are needed only in cases where human exposure exceeds certain values. The difficulty, however, of making reliable estimates of the potential for these deposits to be transferred has greatly hampered the development of the subject and the wide establishment of formal regulatory requirements.

APPRAISAL OF HAZARD: INTERPRETATION OF DATA

Most authorities make their primary appraisal of pesticide hazard on the basis of acute toxicological effects and likely exposure. There is much to be said for this approach because nearly all pesticide accidents do stem from acute effects. As will be seen later, acute data, by virtue of their numerical character, lend themselves to classification systems, which form the basis of so many national and international appraisal schemes.

One of the first issues to consider is the relative importance of data developed from the oral and dermal routes. In most outdoor applications, the commonest source of exposure is contamination of the skin, so that dermal rather than oral data are the more relevant for the assessment of hazard. And yet, curiously enough, many important authorities allow oral data to weigh the more heavily in their appraisal. Admittedly, oral data are the more relevant in the case of accidental ingestion or indeed suicide, but ingestion is not the primary route of exposure that gives rise to health hazards in the field.

The second point to stress is that the acute data can be reliably interpreted only if they refer to the formulation to be sold, or at least, a preparation that is very close to it. For highly toxic products formulated with adjuvants of relatively low toxicity, the oral toxicity of the formulation can be inferred to some extent by extrapolation, but dermal toxicity is strongly influenced by the type of formulation employed and by the ingredients in it. To take an extreme example, the dermal toxicity of a 10% active ingredient granule could be an order of magnitude less than that of a 10% emulsifiable concentrate. Similarly, the dermal toxicity of some solvent-based products is markedly affected by the nature of the solvent, primarily because the solvent can determine the ease with which the active ingredient penetrates the skin.

The appraisal of inhalational hazard also deserves comment. Inhalational hazard is mainly relevant in the case of vapours, particularly when pesticides are applied in an enclosed space, and possibly in the case of fine sprays from ULV equipment or dust, but there are some authorities who attach importance to inhalational hazard even in the case of products of very low volatility applied outdoors as sprays too coarse to be inhaled. Since the data on which such appraisals have to be based are developed in circumstances where the exposure of the animals is almost certainly predominantly dermal anyhow, in spite of efforts to avoid this complication, it is very difficult to appreciate the validity of the appraisal in such cases.

In addition to the numerical features of the data, there are several other very important aspects of the acute toxicity of a product which need to be taken into account when appraising hazard. The Council of Europe (1984) gives a very timely reminder that the temptation to appraise hazard solely on numbers should be resisted. Thus, such factors as delay in the onset of toxic symptoms, slow reversion to normality after effects have occurred, or the lack of an effective antidote, are examples of the more qualitative factors that should also be given due weight in the appraisal.

In addition to information on the acute effects of the product, mention has already been made of the importance that attaches to longer-term toxicology, developed on the active ingredient. The professional applicator may well be subject to small but long-term exposures during his occupation and it is obvious that substances, which at low dosages pose, for example, a carcinogenic hazard to man would have to be treated with great reserve. The science of the interpretation of long-term toxicological effects is, of course, a subject in its own right and the importance for occupational safety of pesticides is discussed further in Chapter 7.

CLASSIFICATION BY TOXICITY

With the object of assisting authorities in their decisions, several countries have introduced schemes for the classification of pesticides according to numerical values for their acute toxicity. Although this approach excludes some important aspects it has become a very popular one. At the present time, there is a considerable disparity between the schemes operated by different countries. Some authorities favour schemes based on active ingredients; some on formulations, and some on a mixture of both. Some feel that oral toxicity should be the main criterion, whilst others recognise the comparatively greater importance of dermal toxicity. Some schemes are very rigidly numerical whilst others allow judgements to be modified in the face of some of the more qualitative aspects of the toxicological properties of the product.

In 1975 the World Health Organisation (WHO, 1975) published a proposed scheme for the classification of pesticides, based on the following principles.

(i) The basis is the acute oral toxicity to the rat, unless the product is conspicuously more toxic to another species or if there is evidence that data from a species other than the rat are more relevant to man.

(ii) Dermal toxicity is also taken into account if the dermal data show the product to be especially hazardous by the dermal route.

(iii) The classification assigned to the product can be amended if there is evidence that the acute toxicity to man differs from that indicated by the animal data alone.

(iv) The scheme is intended to be applied to the formulation as sold. In the absence of data specific to the formulation, figures may be extrapolated from those referring to the active ingredient(s). The more commonly used active ingredients are classified, by way of example, in an appendix to the latest revision of the scheme (WHO, 1982). There is provision for these classifications to be amended if new data so require.

(v) Specific label warnings are suggested for those products that fall into the more toxic classes.

In spite of the occasional anomalies - and it is difficult to devise a numerical system without any - the scheme has been widely accepted, at least in principle. In the latest revision (WHO, 1982) there is provision for toxicological data additional to the LD_{50} figures also to be considered in deciding the classification for a given product. The basis for the scheme is shown in Table 6.1.

In this table, the terms liquid or solid refer to the products being classified. Gaseous or volatile fumigants are not classified under the scheme nor are criteria put forward for maximum allowable air levels. There is, however, a list of volatile fumigants cited and attention is drawn to various national criteria.

In Western Europe, the Council of Europe has endorsed the scheme (1984) and the EEC has introduced a scheme for Member States similar to the WHO scheme. The original EEC proposal was put forward in 1978 as the Classification, Packaging and

TABLE 6.1 WHO Classification Criteria

CLASS		RAT LD_{50} (mg/kg bodyweight)			
		ORAL		DERMAL	
		SOLIDS	LIQUIDS	SOLIDS	LIQUIDS
Ia	Extremely hazardous	5 or less	20 or less	10 or less	40 or less
Ib	Highly hazardous	5-50	20-200	10-100	40-400
II	Moderately hazardous	50-500	200-2000	100-1000	400-4000
III	Slightly hazardous	>500	>2000	>1000	>4000

Labelling of Dangerous Preparations (Pesticides) Directive, (1978) and whereas some member states such as Great Britain and Denmark have already implemented this Directive, the majority are awaiting the formal implementation date of October 1985 before doing so. The numerical values differ slightly from those of the WHO scheme and there are only three classes. The basis is shown in Table 6.2.

The EEC scheme does admit the possibility of more general toxicological data being taken into account and in that sense too it parallels the WHO scheme. Unlike the WHO scheme, it does classify products that occur in gaseous or vapour form; the basis is the inhalational four-hour LC_{50} for the rat. Products with an LC_{50} below 0.5 mg/litre of air are regarded as "very toxic"; 0.5-2.0 mg/litre as "toxic" and 2.0-20 mg/litre as "harmful".

As in the case of the WHO scheme, the EEC scheme explains how to estimate the toxicity of an unknown formulation from data referring to the active ingredient but, unlike the WHO scheme, it lists a series of active ingredients in the "very toxic" class as "Non Transferable". Products containing them have to remain in the "very toxic" class, even though the toxicity of the formulation would otherwise have allowed it to be classified in a less toxic class.

Two examples of the way that member states have adopted the EEC scheme are Denmark (1980) and the UK (PSPS, 1983). In both of these countries, the numerical system is the same as that proposed by the EEC, but the concept of non-transferability of the more toxic active ingredients has not been taken up.

It is important to recognise that in the UK, the EEC

TABLE 6.2 EEC Classification Scheme

CLASS	RAT LD_{50} (mg/kg bodyweight)			
	ORAL		DERMAL	
	SOLIDS	LIQUIDS	SOLIDS	LIQUIDS
Very toxic	5 or less	25 or less	10 or less	50 or less
Toxic	5-50	25-200	10-100	50-400
Harmful	50-500	200-2000	100-1000	400-4000

classification is embodied in the Pesticides Safety Precautions Scheme which is not statutory although because of a series of legislative problems that have arisen under the Treaty of Rome, steps are being taken to make it so in the not too distant future. There are statutory provisions in the UK however which fall under the Poisonous Substances in Agricultural Regulations 1984, which themselves fall under the wider provisions of the Health and Safety at Work Act 1974. These regulations cover pesticide formulations containing upwards of some 60 active ingredients which are described as "Scheduled" products which have been classified essentially on the numerical basis of acute toxicity.

Whilst the WHO and EEC schemes have achieved considerable acceptance in their respective spheres, some countries still operate numerical schemes which are appreciably different. Thus, in the United States, the scheme operated by the Environment Protection Agency (40 CFR/62) is based on four groups, themselves distinguished on the basis of the acute toxicity of the formulation concerned. The steps, expressed in the normal units for the acute oral toxicity to the rat, are 0-50, 50-500, 500-5000 and greater than 5000. The corresponding dermal figures are 0-200, 200-2000, 2000-20,000 and greater than 20,000. There is no distinction between liquids and solids. It should be remembered that in addition to this scheme, there is provision under FIFRA for pesticides to be assigned to a "Restricted Use" category. This is over and above the numerical scheme and takes into account special toxicological characteristics as well as, in some instances, environmental effects.

The classification scheme in Brazil also depends on the rat oral LD_{50} (Brazil, 1980). The most toxic class includes all solids with an oral rat LD_{50} below 100. The corresponding figure for liquids is 200. Dermal figures at double these levels must also be satisfied, so that these rather than the oral data are often the limiting factor. The lowest toxicity class requires very high numerical figures which are also a feature of the system employed in Bangladesh (undated paper). These very high figures may sound attractive from a cosmetic standpoint but there is no hint as to how they are intended to be determined; it is normally impracticable to dose a rat with more than 2000 mg/kg for the determination of acute dermal toxicity. In the US system, the highest dermal figure mentioned for a liquid is, when projected onto an average human bodyweight, equivalent to something in the region of half a litre, far more than the normal body can retain as a skin contamination.

A feature worth noting in the Brazilian scheme is that products containing active ingredients of very high toxicity are placed in the most toxic class, irrespective of the actual toxicity of the formulation; the critical figure is an oral rat LD_{50} below 25 mg/kg bodyweight, rather akin to the non-transferable concept in the EEC scheme. Like the EEC scheme, the Brazilian classification also covers vapours (using one-hour rat LC_{50} data as the criteria). It also requires that products considered to represent a carcinogenic, mutagenic or teratogenic hazard to man automatically fall into the most toxic category.

The proposed classification scheme in Japan (MAFF, 1983) (Table 6.3) is a numerical one but unlike most others it is based primarily on the active ingredient. Formulation limits are cited for the more toxic categories and since these are all ten times higher than the active ingredient figure, it is they rather than the active ingredient limits that will usually be determinant. The scheme does not cite the rat as test animal; the required data guidelines specify two species of test animals so that presumably the lower of the two figures is used as the basis. Dermal as well as oral data are taken into account and since the dermal limits are so very much higher than the oral ones, the scheme does appear to recognise the relatively greater importance of dermal data. However the need for subcutaneous and intraperitoneal route data is unusual.

Table 6.3, however, is only a part of the scheme. As in many other numerical systems, additional judgements can also be applied by taking into account a range of long-term toxicological effects, some of which can prevent the registration of the product entirely. A particularly serious problem arises in the case of pesticides of high toxicity and no specific antidote.

The classification system in operation in Canada as described by Franklin and Muir (1982) is particularly

TABLE 6.3 Proposed Japanese Toxicity Classification Scheme

PARAMETER	CLASS					
	I	II	III	IV	V	VI
Oral LD_{50} a.i.	<3	<30	<300	<1500	<3000	>3000
Oral LD_{50} formulation	<30	<300	<3000	<500	>5000	
Dermal LD_{50} a.i.	<100	<1000	<5000			
Dermal LD_{50} formulation	<1000	<10,000				
Inhalation LC_{50} a.i.	<200	<2000				
Subcutaneous LD_{50} a.i.	<20	<200				
Intraperitoneal LD_{50} a.i.	<10	<100				

interesting because it does not appear as dependent on numerical classification as are most other well known systems. Moreover, the classification depends not only on the toxicity as such but on the use to which the product will be put.

At the federal level, pesticides are classified as Restricted, Commercial or Domestic. The classification is made according to the requirements of the Canadian Pest Control Products Act. Products intended for application to environmentally sensitive areas are more likely to be classified as restricted; so much so that all pesticides, irrespective of their toxicity, are classified as restricted, when used for the treatment of forests or aquatic environments.

The position in Canada is complicated by the fact that, as well as the federal authorities, provincial governments may also apply their own classification by placing the product into a "schedule"; there are usually 5-6 where separate provincial schemes are established. In general there would appear to be a reasonable degree of co-ordination between the federal and provincial classifications, although there are frequently substantial differences of detail.

Although provincial involvement is a complication there

is much to be said for the Canadian system. Through not being as rigidly bound to a numerical system the authority has greater scope for judgement and the scheme is not plagued by the uncertainties that always attach to products with a toxicity falling on the borderline between two categories. Moreover, it allows the authority to make certain that the treatment of environmentally sensitive areas is undertaken only by skilled people.

It is indeed greatly to be hoped that, in the future evolution of classification schemes, there will be enough scope for realistic appraisals always to have precedence over mere numerical conformity.

IMPLEMENTATION

Having appraised the toxic properties of the pesticide product, whether by means of a numerical classification system or by more general criteria, the control authority has to decide what measures need to be taken for the product to be placed on the market. Unacceptable risks, seen in the context of its national circumstances, must be avoided. Obviously, circumstances vary from country to country so that different authorities can and do take different decisions on a given product. There are two main issues to be considered from a purely regulatory standpoint: presentation of the product, mainly labelling and packaging, and the need for restrictions on availability.

PRODUCT PRESENTATION: LABELLING

The label on the package is one of the most important, if not the only channel of communication between the manufacturer and the user. Many national and international authorities have devoted much thought to the derivation of the ideal label. Thus, the 2nd FAO Intergovernment Consultation on the International Harmonisation of Pesticide Registration Requirements (FAO, 1982) concluded that "Although registration procedures for pesticides have become increasingly complex, the end-product of registration has remained the same - a labelled package containing a pesticide formulation ... labelling is the main method of identifying the product and of communicating instructions and advice to all concerned with its handling". The meeting went on to say that "For successful communication a label must be easy to read and understand and that printed and graphic material ... should bear all the necessary information and instructions for effective and safe use in a language understood by the user". Without doubt, the most widely advocated advice by both official agencies and industry is "READ THE LABEL".

Putting forward its own Guidelines on Good Labelling Practice for Pesticides, the meeting strongly recommended the maximum possible degree of harmonisation. It made recommendations on precautionary statements which many authorities now accept. Those aspects dealing with operator safety directly were summarised as follows.

> "Appropriate and clear indications of the degree and the type of hazard, using the relevant warning of risk symbols, should appear on the label, when the nature of the formulated product makes this necessary. These should be in keeping with a harmonised classification of pesticides by hazard, preferably that proposed by WHO.
>
> Appropriate instructions, in the form of standardised safety phrases, for the protection of consumers, operators, livestock, domestic animals, wildlife and the third parties.
>
> The recommended first aid, antidote (if any) and other information for physicians as required by the appropriate health authorities, when the toxicity of the formulated product warrants it."

The Intergovernmental Consultation further recommended the adoption of standard phrases where possible, a point which forms a feature of the EEC Council Directive relating to the Classification, Packaging and Labelling of Dangerous Preparations (Pesticides) (1978). Whilst the Directive goes into considerable detail on such issues as the size of labels for different sized containers, appropriate warning symbols, and language, one of its most important provisions is a series of standard phrases. These are intended to be selected by the competent authority to appear on the label according to the toxicity classification into which the product falls. They are divided into "Risk" phrases (e.g. "Very toxic by inhalation") and "Safety" phrases (e.g. "Do not breathe dust"). There are over thirty phrases in the list. The value of this approach, of course, is to encourage harmonisation between member states who are required to enact domestic regulations which put the Directive into effect. Symbolic indications of the nature of risk are also required, again according to the classification assigned to the pesticide. These are specified in an earlier Directive (67/548/EEC, Annex V). These same symbols are detailed in the British Working

Document E2 (PSPS, 1983) which requires the EEC standard risk and safety phrases to be used according to the toxicity class of the product. The Council of Europe (1984) also supports the principle of using standard phrases although its list is considerably more extensive. As is pointed out in the text, the value of the use of standard phrases is to promote the uniformity of warnings on the labels of different formulations posing similar risks. The FAO Intergovernmental Consultation also supports a standard phrase approach in its Labelling Guidelines (FAO, 1982).

In the United States, labelling requirements are somewhat less specific. As in most other countries, the label proposed by the manufacturer must always be agreed with the registration authority. Very toxic products, as defined by the EPA classification system, must bear the skull and cross bones symbol and the word "Poison" (in red) in a prominent position on a background of contrasting colour, together with a statement of a practical treatment to be administered in cases of poisoning.

Labelling is a subject where, more than most, different national authorities require compliance with very detailed requirements. Since, in many countries, this is a legal obligation, individual regulations must be studied with great care by the manufacturer and in most cases labels must be submitted in draft for approval. Nevertheless, views on labelling are bound to continue their evolution as experience develops, and by way of a summary of the main aspects normally reflected in the regulations from most advanced countries, the following points need always to be borne in mind.

(i) Few people enjoy reading safety precautions. There is always a temptation for regulations to require too much to be put onto a label. A label is too small a place to say all there is to be said about a pesticide and overcrowding is a sure way of discouraging the reader.

(ii) Care should also be taken to avoid mere disclaimer advice which makes no contribution to safe use. Such phrases as "Caution - This product has been shown to cause tumours in experimental animals" leaves the reader completely in the air and contributes nothing of real value.

(iii) Space given to negative recommendations should be drastically rationed. Whilst there may be some well accepted cases where such advice is of evident value (e.g. do not breathe the dust), the label should be devoted to explaining what one can do rather that what one cannot. Procedures not recommended on the label should be understood as being unacceptable. Perhaps the one piece of negative advice that is acceptable, therefore, is "This product should not be used for any other purposes than those advised or in any other manner than is advised on this label".

(iv) Maintaining easy legibility is an ever present challenge. As well as the avoidance of unnecessary material, such simple factors as writing large enough to be seen easily, labels resistant to damage, and instructions in an appropriate language are of vital importance. The use of pictograms as an aid to comprehension is an interesting current development.

(v) Use precautions must always be reasonable and practicable. If the advice from national authorities fails on either of these two points, credibility is lost. Thus, advice to wear protective clothing in hot climates, where the applicator might run the risk of heat stroke as a result, is clearly unacceptable. Similarly, trying to play safe by the recommendation of unnecessarily heroic precautions is a sure way of losing credibility, with the result that essential measures for really toxic products run the risk of being ignored.

It is greatly to be hoped that in future evolution of the design of pesticide labels, these points, which have arisen from long experience, will continue to be taken as basic essentials.

PRODUCT PRESENTATION: PACKAGING

Few registration authorities dictate precise details of packaging of pesticides but all nevertheless expect packaging to be of a reasonable and acceptable standard. There are certain rather obvious requirements for packages to meet. The EEC Classification, Packaging and Labelling of Dangerous Preparations (Pesticides) Directive (1978) gives a useful list of basic needs. In essence, packages must be secure, not liable to interact with their contents, strong enough to withstand normal handling and fitted, where appropriate, with tamper-proof closures that can be reclosed after the package has been opened.

Sometimes a package can be adapted to limit availability of the product. Thus, marketing in large drums can reduce the risk of the product reaching, for example, the domestic market. A requirement of this sort is not often a regulatory requirement; packaging is usually agreed between the manufacturer and the authority on a case-by-case basis. According to Palacio (1982), an interesting exception is Colombia where marketing of products that are considered too toxic for use by domestic or ill-informed users must not be sold in packages of less than one litre. He reported, however, that this objective is not always achieved and that abuses are frequent. In some countries, however, limiting the size of the package can be a very effective means of reducing availability to unauthorised persons provided that effective measures are taken to prevent unauthorised repacking into inadequate small packs which may not be properly labelled.

RESTRICTION OF AVAILABILITY

Many authorities feel that the only way to protect users is to restrict the availability of the more toxic products to people of proven ability, or to classes of people more likely to respect use precautions. Thus, some countries have evolved very advanced schemes involving the establishment of operator licencing, but there is great variation between countries and often between states within a country. Of the many countries that have such schemes, the following are selected by way of example as covering the main aspects of what has been achieved, and to illustrate the differences that arise from differing national circumstances.

In Spain (Repetto et al., 1982), where pesticides are classified into five toxic classes, the most toxic are only permitted to be handled by an organisation whose director has a university degree and whose staff are "well qualified". To have these products applied to his crops, the individual farmer must sign a contract with the firm making the application. They in turn must not allow men under the age of 18 years to make the applications and all applicators must be medically examined every six months. There is also a series of restrictive regulations covering the next class, although these are somewhat less severe. The authors regarded the scheme as being relatively successful, but stressed the importance of maintaining good educational programmes and medical monitoring of workers handling the more toxic products. Even so, they considered that much remains to be done to extend the educational effort before there is an acceptable awareness of the importance of adhering to protective and precautionary measures, and even of the legal obligations involved in some cases.

In Canada, according to Franklin and Muir (1982) the decision on who is permitted to use products in the three categories (Restricted, Commercial or Domestic) ultimately rests with the provincial government authorities but, in most provinces, restricted pesticides can only be applied by farmers or professional applicators. In most provinces, there are arrangements for professional applicators to be licenced, although specific arrangements are still to be finalised in two provinces. In some provinces, professional licences are specific to a given use and an applicator may have several licences which each qualify him for a specific aspect of pesticide application. Thus, in the Province of British Colombia, separate licences are issued for ten different classes of pesticide usage, whereas in Newfoundland, with very few farms, a single licence for a general pesticide applicator covers all uses. Procedures for granting licences also vary considerably between different provinces, but most require evidence that applicators have received officially approved training. Whilst in some provinces, sales of very toxic products are confined to licenced applicators, in others individual farmers can also buy and apply to their own land, even though they have no formal training. Franklin and Muir felt it possible that this represents a serious loophole.

As mentioned in the preceding section of this chapter, the Federal Insecticide, Fungicide and Rodenticide Act of the United States (FIFRA) (EPA, 1978) requires that products be classified according to hazard; a product in the "Restricted" class can be applied only by certified applicators. The Administrator of the Environmental Protection Agency (EPA) has to be satisfied that the applicator meets his requirements. It is possible for the Administrator to delegate this responsibility to an individual State to administer, provided that he can be satisfied with the arrangements established by the state authorities. It is apparent that the individual states in the USA have somewhat less authority than the provinces in Canada, but the relation between the federal and state authorities in the USA is complex and care should be taken to avoid facile conclusions. Some amplification of the means by which a state achieves "primacy", i.e. delegated authority, in the enforcement of FIFRA, may be derived by consulting the relevant US Federal Register (1983) item.

The system operated in the UK stems essentially from the obligations imposed by the Poisonous Substances in Agriculture Regulations 1984. These obligations, unlike the requirements of the PSPS are statutory and are imposed not only on employers but on workers and on the self-employed. The scheme does not depend on certification of applicators as is the case in North America. Instead there are a series of very specific precautionary measures that have to be taken by all concerned in the application (HMSO, 1978). Whilst many of the requirements are the sort of precaution the good applicator would take when using any pesticide, the requirements have the force of law only when products containing scheduled substances are being applied. Products containing these substances are not normally available to domestic users.

The regulations are enforced by a series of inspectors appointed under the Health and Safety at Work Act and who have the duty of ensuring that the provisions of health and safety legislation relating to agriculture, including these specific regulations, are properly observed. The inspectors must be notified of any suspected injury to health arising from a scheduled substance since they may be able to give vital help if promptly informed. It is also the role of the inspector to give help and advice as to how the regulations should be put into effect.

An interesting feature of the UK scheme is the recognition that enforcement officers have an important role to assist and advise as well as merely to enforce. Indeed the regulations even make provision for growers to request permission to use alternative precautions, where it can be shown that these are more practicable in the circumstances concerned. This capacity for evolution is an important feature of the scheme, which, in practice, has proved to be very successful.

FUTURE DEVELOPMENTS

There is little doubt that the first requirement for the future is the extension of sound legislative principles to the less developed countries which is where health hazards from pesticides are most likely to arise. In general, the accident record of the developed world is very good and there is little doubt that, in competent hands, pesticides of high toxicity can be applied safely, provided that they have passed through a competent registration process. There is more to safety than legislation, of course. There must be a will to handle pesticides competently in the first place, and then there must be knowledge particularly for the distributors who advise the many small users. Training provides the knowledge and must cover effective and safe use, safety in transport and storage, and safe disposal of empty containers. The introduction of regulations is to provide a back-up to these essential pre-conditions. This means that regulations must be constructive, they must be practical and above all, they must be enforceable. Invidious situations can arise when only the more serious members of the community feel the need to abide by the provisions of the prevailing regulations.

Attempts to introduce unenforceable legislation are bound

to be frustrated. The extension of regulations into countries without them can only proceed successfully where the majority have the ability and the will to support them. Progress in some is bound to be slow, and as an interim measure, the FAO Intergovernmental Consultation (FAO, 1982) recommended that, as a starting point, countries concerned should consider restricting the availability of the more toxic products, using the WHO classification scheme as a basis. In addition, it recommended that consideration be given to the preparation of a code of conduct so that all concerned, especially suppliers, would have guidance on the restrictions that ought to apply to the more toxic products. This code is likely to play a useful role. It is always difficult, of course, to impose constraints on a country from outside and codes tend to be followed only by the more responsible, from whom, in any case, the problems are least likely to originate. One must hope that the constraints that are finally decided do not merely make life more difficult for the serious manufacturer and user without playing a significant role in the protection of the health of the applicator.

REFERENCES

Brazil (1980) Decree No 04/DISAD 30.4.80. Official Order of 6th May 1980.

Bangladesh (undated paper) Standards for Selection of Pesticides for Registration in Bangladesh.

Canada (1978) Draft Registration Guidelines.

Classification, Packaging and Labelling of Dangerous Preparations (Pesticides) Directive (1978) Off. J. European Communities 29 July 1978 L206/13-L206/25.

Council of Europe (1984) Pesticides, 6th Edition, Strasbourg.

Denmark (1980) Statutory Order No 410, Pesticides in Denmark.

EPA, (40 CFR/62) Registration, Re-registration and Classification Procedures Section 162.10 - Labelling Requirements. US Code of Federal Regulations Title 40 Part 162 Subpart A.

EPA (1978) The Federal Insecticides Fungicides and Rodenticide Act as Amended 1978. US Environmental Protection Agency Nov. 1978 OPA 17/9.

EPA (1982a) Pesticide Assessment Guidelines - Subdivision F - Hazard Evaluation: Human and Domestic Animals. US National Technical Information Service.

EPA (1982b) Pesticide Assessment Guidelines - Subdivision K - Exposure Data Requirements: Reentry protection. US National Technical Information Service.

FAO (1982) Report of the Second Governmental Consultation on International Harmonisation of Pesticide Registration Requirements (AGP 1982/M/5).

Franklin, C.A. and Muir, N.J., (1982) Education of Licensing Procedures for Pesticide Applicators and Vendors in Canada. Studies in Environmental Science, 18, 89-103.

GIFAP (1983) Guidelines for the Safe and Effective Use of Pesticides, GIFAP, Brussels.

HMSO (1978) Poisonous Chemicals on the Farm. London.

Jeyaratnam, J.J. (1982) Health Hazards Awareness of Pesticide Applicators. Studies in Environmental Science, 18, 23-30.

MAFF (1983) Ministry of Agriculture Fisheries and Forestry, Japan, Proposed Guidelines for the Evaluation of the Safety in Use of an Agricultural Pesticide.

Palacio, D.C. (1982) Accident Prevention and Education for Safe Use of Pesticides in Colombia. Studies in Environmental Science, 18, 137-146.

Popendorf, W.J. and Leffingwell, J.T. (1982) Regulation of OP pesticide residues for farm worker protection. Residue Reviews, 82, 125-201

PSPS (1983) UK Pesticides Safety Precautions Scheme.

Repetto, M. et al (1982) Prevention in the Use and Applications of Pesticides in Spain. Studies in Environmental Science, 18, 161-69.

US Federal Register (1983) 48, 404-411. FIFRA State Primary Enforcement Responsibilities.

WHO (1975) Recommended Classification of Pesticide by the Hazard. WHO Chronicle, 29, 397-401.

WHO (1982) Guidelines for the Use of the WHO Recommended Classification of Pesticides by Hazard. WHO letter VCB/78.1 Rev. 3.

CHAPTER 7

CURRENT TRENDS AND FUTURE NEEDS

G.J. Turnbull

INTRODUCTION

Workers handling pesticides in agriculture and public health are not suffering from an epidemic of poisoning due to their work. Adverse health effects, either acute poisoning or delayed effects like cancer, never have been major occupational threats to health and welfare. The least developed, the developing and the developed countries are alike in this respect. In countries with reliable statistics it is evident that injuries to workers caused by pesticides are uncommon compared with those caused by working on farms with machinery, or injuries caused by falls, lifting excessive weights or manual cultivation. A sense of proportion is needed when considering the occupational hazards of pesticide use. Beyond all doubt, the beneficial effects of pesticides outweigh the risks. Nevertheless, there is no reason for complacency as even one case of work-related accidental injury or harm to health is one too many, and undoubtedly there is still room for improvement.

Misuse is by far the most frequent cause of reported harm, but there have been complaints of harmful effects on health from even the recommended use of pesticides in several countries (Espir et al., 1970; Hardell and Sandstrom, 1979; Kilgore and Akesson, 1980; Reeves et al., 1981; International Organisation of Consumers Union/Friends of the Earth Malaysia, 1982; Balarajan and McDowall, 1983; Blair et al., 1983; Caufield, 1984; Friends of the Earth, 1984). The evidence in support of these complaints so far has been poor and has not been confirmed by formal investigation. The conclusion of many governments, international bodies and independent research workers is that the recommended uses of pesticides do not produce adverse health effects in man (WHO Data Sheets on Pesticides, 1975-1984; IARC, 1979 and 1983; Vetorazzi, 1979; Advisory Committee on Pesticides, 1980; Murphy, 1980; WHO, 1981; FAO, 1982; Hayes, 1982; Golding and Sladden, 1983).

Adverse health effects from pesticides have occurred, however, in factories manufacturing and packing the products prior to sale. In the course of normal work to produce the nematicide dibromochloropropane in the USA some workers suffered contamination which caused them impaired fertility (Whorton et al., 1971; Anon., 1978). In fact the toxicity of that substance had long been known and the problem was not a lack of published hazard information (Torkelson et al., 1961). Such instances are, thankfully, not at all common. As in other major industries, there have been major accidents involving leaks, spills, fires or explosions in pesticide manufacturing works and in distribution and storage. The largest accident of this kind was at Bhopal in India in 1984 when there was a massive death toll. It would not be fair to conclude, though, that the manufacture of pesticides is intrinsically less safe than, for example, the petrochemical or mining industries.

Factual information on the safety record of pesticides reveals a problem in the least developed and developing countries. In some countries, many of the poisonings that do result from pesticides are suicides and are not due to malpractice at work or to excessive pesticide exposure in farming. In spite of the work that has been done in individual countries there is a serious lack of information on the extent to which people are said to be harmed and why. Assumptions are widespread but proper investigation of cases is so infrequent in some areas that the factual information is sparse and incomplete. Consequently, pesticides probably are blamed for many illnesses in the developing countries which, upon investigation, would be found to have other causes. The attitude in the developed countries also is to blame pesticides for many of the ailments which have no visible cause.

The challenge that industry, governments and international bodies now face is to identify more sharply the commoner causes of poisoning by pesticides in different countries, so that efforts can be concentrated on those issues where the yield will be greatest. For this purpose, we need more information, and we need information from those countries least able to provide it. What are the most common causes of accidental over-exposure to pesticides? They probably vary considerably from country to country, but we do not really know and there might be some surprises. Especially in some regions there is a need to reduce the numbers of suicides involving pesticides.

EXPOSURE STUDIES

For some types of work with pesticides there is considerable information on the likely level of worker exposure during application in agriculture and public health. However, for other work little or nothing appears to be known. For example, huge numbers of cattle and sheep are dipped every year in insecticide to protect the animals' health. Each animal is totally immersed for an instant and there seems to be a dearth of information as to just how much skin contact there can be for the workers. There may be a need to measure the amount of contamination of workers doing such tasks, although it is widely known already what precautions should be taken to avoid personal contamination.

However, it is clear from Chapter 3 that not every pesticide sold in a country should be examined in operator exposure studies for each of its intended uses. The earlier chapters showed how the likely exposure could usually be calculated from data already obtained during the study of other products in comparable uses.

In fact, with the limited funds available to national authorities, especially in the developing and least developed countries it may often be best to give most support to training programmes. The benefit of better training will often be far greater than the information gained in local worker exposure studies or, for that matter, local analysis of pesticide residues in food crops.

The remedy for non-occupational poisoning (including suicide) in least developed and developing countries is not, in fact, enormously expensive. It lies in the areas of training, the establishment of so-called good agricultural practices including safe storage, and in the communication of manufacturers' hazard information to work supervisors and to the medical centres.

USING EXPOSURE DATA

Previous chapters described how information on worker exposure to one pesticide can often be employed to estimate the exposure to another pesticide used in a similar manner. A distinction has to be drawn, however, between the estimated actual exposure from contamination of the skin and the theoretical potential exposure of the body surface including clothes and protective gloves as well as the bare skin of, for example the face or arms. Since most of the potential contamination is of the hand area the precaution of wearing gloves at appropriate times drastically reduces the ultimate contamination. Also, the potential contamination of other parts of the body generally involves areas normally covered by working clothes. There is no constant relationship between potential and the actual (lower) skin contamination since both the type of clothing worn, such as short or long sleeves, and the requirement to wear gloves are variable. Generally though, the information on potential exposure presented in this book shows the worst likely exposure as well as the range that may arise in the course of typical working. Usually the actual exposure will be in the lower range of values provided.

Better use could be made of the information already available on potential skin exposure. The problem is two fold. There is a lack of published information on the sort of clothing actually worn in agriculture and on the pattern of work, particularly the time spent in a day, and over successive weeks, actually spraying and handling pesticides. Assumptions can be made but there remains a difficulty in calculating a reliable answer to the chain of "what if" questions that are part of any evaluation of exposure. Knowing the measured exposure, what if the workers actually wore short rather than long-sleeved shirts, what if they wore no gloves for a particular task, what if the work clothes were penetrated by a certain proportion of the chemical contamination, what if there were secondary transfer from, say, gloves to the skin of the face, what if the workers washed the exposed skin of the face and hands at every refilling rather than at major breaks, what if the chemical were particularly well, or very poorly absorbed from the skin, and so on.

There is quite extensive and detailed information on many of these aspects but the problem is to search out the relevant data in order to draw comparisons and make the calculations. Present day information technology would enable that to be done more rapidly and more comprehensively if all the relevant information were entered on a computer data base including details which affect subsequent calculations. If that were done government, industry and other bodies would be better able to judge the degree and importance of pesticide exposure in specific circumstances.

By knowing the quantitative effect of alternative work practices and of individual variation in, for example, absorption of chemical from the skin it would be possible to make two important advances. Firstly, the observations, analyses and calculations could be checked by making comparisons with the quantitative information on the excretion of pesticides by workers after normal work involving exposure. Secondly, it would open up the mass of information to many more people and allow more complete and reasoned answers to be given to all those with a concern about worker exposure. The benefits of such a computerised data base are evident, as is the need for it. A national or international agency is better placed to do this than the pesticide manufacturing industry.

HAZARD LABELLING OF PESTICIDE CONTAINERS

A pesticide stored in the wrong container or without a label giving appropriate warnings is likely to be a hazard, especially to childen. Too many accidental poisonings involve children who should never have had access to the material. There is a great need to prevent transfer of pesticide from the manufacturer's containers (discussed in Chapter 6) to unlabelled domestic bottles whose contents may then be drunk mistakenly, particularly by children.

Hazard symbols and safety instructions can only be effective if they are seen and the advice followed. The label is the principal and most immediate means of conveying vital information on safe working. But many labels giving safety instructions are dull, overcrowded, and even confusing to read due to ponderous official phrases and rigidly controlled presentation. Perhaps it is only the high cost of the chemical that encourages workers, and perhaps only the supervisors, to read the label. Both industry and government authorities have put great effort into the careful labelling of pesticides (see for example Gusman and Irwin, 1980; Jellinek, 1984; Lirtzman, 1984). However, the amount and layout of information on the label often reduces comprehension so that the impact of the safety precautions is insufficient in view of their importance. Surely there is scope for innovation here. The need is constantly to campaign for labels to be read and the instructions followed closely. If there is a lack of literacy or if the local language or dialect has not been used then good pictograms promise to play a valuable role.

A 1978 workshop in the USA and an FAO publication both highlighted some of the guiding principles for constructing labels (Gusman and Irwin, 1980; FAO, 1983a). Above all, a label that is complex also is confusing. Text and symbols that warn of hazards associated with a product should be easy to understand and difficult to misunderstand. A few special symbols to warn of a limited number of generic hazards are the best means of communicating a warning alert and identification of the hazard. The text of the label can then convey additional important information such as the precautions which minimise the hazard. A person reading the label will respond to a hazard warning in a way that reflects previous experience, hence the need for training of farm workers to include their being taught to look at the label.

Many pesticide products sold to farmers are less likely to poison a person than many household cleaning agents, petrol or paint solvents or even some alcoholic beverages. The lack of a warning symbol on a label for such pesticides does not mean that the message on the label is unimportant.

It would be quite wrong to label such pesticides with uncalled for warning symbols that were, in essence, false alarms because attention would then be diverted from the products which deserve the hazard warning. If an alarm message is given with every product then, after a while, and even with proper training, the alarm message is simply not seen; it becomes too familiar. Restrictions on the choice of pesticides available will not force farm workers to act on the label instructions for hazard precautions. Neither is there benefit from absolute standardisation of the size, shape or colour of containers for pesticides. The labels on pesticide containers are already highly standardised. The claim that fewer products for farmers to choose from would mean fewer accidents is illusory. Bans, restrictions on use and curtailed choice are not the means to improve the way pesticides are stored and used.

TRAINING

The ability of the person using pesticides to work according to the label instructions is evident from his preparations even before the container of pesticide is opened. Only if a reasonable amount of appropriate training and supervision is given to each operator can satisfactory working methods be expected. Training has to be realistic and relevant and, particularly for safety issues, it should be repeated periodically. Lack of proper training can only result in greater worker exposure which is needless and in extreme cases perhaps hazardous. Safe systems of work go hand in hand with training. Potential hazards need to be appreciated so that they can be avoided.

It will perhaps be argued that in certain parts of some countries there is too little trained manpower to pass that training adequately on to the general labour force. The best deployment of the trained manpower may then be as specialist teams handling the application of certain pesticides which are classified as more toxic than the others. Such spray teams are already used in national malaria control programmes involving house treatment with insecticide to kill adult mosquitos. Meantime the "Train the Trainer" approach is an efficient way of injecting skills into countries needing advice.

A major objective of the pesticide manufacturer is to continue to provide the means for effective crop and public health protection without injury to the health of those concerned. The industry has come a long way in the pursuit of those aims. It has developed better products; it has improved formulations, packages and labels; it has collaborated with officials at both the national and international level to improve safety measures, and it has invested more effort in the propagation of safe-use education than is commonly realised. But there is always scope for improvement in training.

CONTAINER DESIGN

Pesticide containers vary from small to large drums, from foil sachets to large bags or kegs. All are intended to survive the rigours of transport without bursting or leaking or causing contamination. However, some containers are difficult to open and so tend to contaminate the hands during opening, especially if a screwdriver is used to pierce the top or if a bag of powder is opened carelessly.

Large containers are difficult to lift and to pour from and some drums inevitably leave a trickle of liquid from the spout and down the side, while transfer of powders without spillage or creation of airborne dust is frequently difficult. This can only bring about contamination. There are some national and international guidelines on container design (FAO, 1983b; Pesticide Container Design Panel, 1984), but surely there is scope for technical innovation to produce pesticide containers that will not leak or burst but will pour properly without contaminating the operator. It is handling of the undiluted pesticide that gives rise to the majority of personal contamination and the design of containers is seen to be a critical area where improvements must surely be possible. A development towards this end was the so-called Bozzle® ready-to-use spray pack for the Electrodyn® spray machine (Coffee, 1981). This is a novel concept in which the chemical is supplied in an integral assembly of disposable plastic nozzle and container.

SPRAYING EQUIPMENT

It is not possible to guarantee that the spray equipment used on farms and in public health work will never leak. It is at least possible to ensure that workers regularly check for leaks during work and are trained to take appropriate action rather than suffer gross contamination with the diluted pesticide.

Unfortunately, it is still possible for contamination to occur with current spray equipment simply during transfer of the undiluted pesticide into the tank of the sprayer. Some of the tractor-powered equipment requires the agility of a gymnast to get the heavy container of pesticide up to the port on top of the tank. A variety of closed systems have been suggested which would transfer pesticide formulations from the original containers to the tank of the sprayer with minimum handling (Jacobs, 1984). However, even when suction probes or wands are provided to transfer liquid from containers at ground level up into the tank of the sprayer there are still opportunities for subsequent personal contamination from the probe.

There is considerable scope for improvement in the design of filling systems, with rinse cycles to clean the empty drums and filling probe, with minimal opportunity for drips and splashes, and without the need to clamber over or handle equipment that is contaminated after the first filling and spraying.

TIME OF WORK

While no one should ever think that sloppy work practices are ever acceptable because only a short period of work is involved restricting the hours of work with a pesticide is one way to reduce the opportunity for over-exposure. This is particularly relevant in agriculture. In large plantations or on large farms it would be possible for one person to do all or most of the spraying and so work with pesticides for periods which in other occupations would be regarded as a very long working day or week. In the developed countries farmers spray at any time from dawn to dusk at certain times of the year. Regulations are therefore framed to restrict work periods with certain specific pesticides to sensible lengths. The British Poisonous Substances in Agriculture Regulations 1984, for example, require workers handling a limited range of the more toxic products to work with them for no more than 7 hours during any day, 40 hours in any seven consecutive days or 80 hours in any 21 days. Such time limitations contribute to avoidance of excessive exposure.

PROTECTIVE CLOTHING

The last line of defence between the chemical and the person is the protective clothing. Appropriate protective clothing must be worn when recommended by the label on the product. This may mean some or all of gloves, boots, coverall, apron and face visor or respirator. However, in tropical climates, heat stress restricts the use of extensive protective garments and the use of heavy materials for clothing (Staiff et al., 1982). The need is to vary the type of pesticide products available or provide more suitable protection.

Many protective garments, including gloves, will sooner or later be penetrated by some chemical. The rate at which this break-through happens can be measured and there is ample evidence that grossly contaminated gloves will protect only for a limited period of time and, indeed, can aggravate contamination instead of protecting against it. It may be possible to choose glove materials with a longer penetration time, but the ultimate safeguard against penetration must be periodical replacement. More detailed guidance from glove manufacturers on this aspect would be helpful.

If protective clothing is worn, for instance gloves or overalls, these must be cleaned at regular intervals and replaced periodically. Judging from the condition of gloves seen in industry and agriculture, quite often more could be done to provide clean protective clothing. For example, where, on a tractor-mounted sprayer of the type commonly used in agriculture in developed countries, should the contaminated rubber gloves be stored between use? Generally there is no sensible place to put them and no water to wash them before removal. Then, of course, the temptation is to put them in the locker with odd tools and parts or, worse still, in the cab.

At the end of a period of work the items of protective clothing presumably are contaminated with pesticide and so they must be taken off with a degree of caution. This applies particularly to gloves which can spread contamination onto clean skin while being taken off. A good lesson is to smear carbon black or similar marker over a pair of protective gloves during wearing, then to remove the gloves in the usual way and see how far the pigment has spread onto clean skin and, worse still, the previously clean inner surface of the gloves. Contaminated gloves, far from providing protection can, all too easily, become a warm moist chemical poultice.

Traditional cleaning procedures for protective garments can be adequate for removal of superficial contamination. In developed countries one way to deal with grossly contaminated protective clothing is to discard the items involved. Otherwise it may be possible to wash fabric garments clean but this has to be done at fairly frequent intervals (Finley et al., 1974). Visual inspection may not always be an adequate check for effective cleaning. Gloves and foot-wear can only be superficially cleaned and any chemical that has penetrated the material will eventually reach the side next to the skin.

SOAP AND WATER

This is the oldest and best established remedy for contamination of skin by chemicals. That is true in locations with water, such as the typical farm yard in England on a mild day in spring, but water is not always made available for personal hygiene. A bowl of water, a piece of soap and a clean towel will be found in the field only if someone makes an effort to provide them. And this is just as true in developed countries as it is in a remote location in a developing country. But these simple supplies are vital. This is such an effective way of dealing with contamination because it removes much of the contamination before it can be absorbed or cause skin irritation. Use of water even without soap and towel is a great improvement on no washing at all. If it is technically possible to transport the pesticide to the place where it is to be applied, whether as a spray, a dust or a fog, then it must surely be technically possible also to transport sufficient water to allow the workers to wash themselves clean of contamination. This should always be done.

CONTAMINATED EQUIPMENT

Once work has started all the equipment in use represents a potential source of contamination. Spray booms, nozzles and hoses are bound to be covered with spray and the careful worker will not handle these with bare hands until they have been cleaned. Less obviously, even the outsides of tanks and drums can also cause contamination particularly as the work progresses. Again, scrupulous cleanliness, good protective clothing where necessary, and avoidance of carelessness are the essential defences against contamination from this source. The benefit of a readily available supply of uncontaminated water is obvious. The problem is to provide water for cleaning equipment where water is scarce. The need is to persuade everyone involved that a good standard of housekeeping and cleanliness of equipment has a tangible benefit by reducing personal exposure to the chemicals.

SCIENCE AND SAFETY EVALUATIONS

Safe use of pesticides is a practical subject. It is helpful to understand the scientific basis for effective safety measures but it is not absolutely essential;

sometimes understanding the science comes after the problem is solved. For chemicals, safety includes avoiding acute poisonings or local irritation from skin or eye contact, and avoiding delayed harm such as cancer. Toxicology, which is the study of poisons, has developed greatly as the number of chemicals approved for everyday use has expanded. By means of model systems, including methods using laboratory animals as surrogates for man, the nature of a chemical's toxicity is identified and then quantified in terms of a dose response relationship. An adverse effect is evaluated for its relevance to man (and other possible target species in the case of pesticides).

Pesticide Safety Evaluations

The science of toxicology has an increasing contribution to make in defining the safe uses of chemicals, supplementing (but not replacing) the empiricism of earlier years. It is the scientific foundation for the safety evaluation of pesticides. The empirical method for evaluating the occupational safety of a pesticide is founded on a large body of experience of pesticide use. Analogies are drawn between the toxic properties of the different pesticides which have a particular use, for example the various cereal fungicides used in Europe. The control procedures and protective measures needed for a new formulation can then be deduced from the practical experience of the measures known to be effective for that use. Additionally, comparisons are made between the measures needed to limit exposure during one use of a specific pesticide and the measures which would be practicable and effective for a quite different use. The effectiveness of this approach in solving the early problems of occupational over-exposure is demonstrated by the fall in number of occupational poisonings from, for example, DNOC, parathion and certain other organophosphorus compounds when precautions were improved, and by the present sustained low level of occupational incidents in the developed countries.

The very comprehensive evaluation of the safety of pesticide residues in food can now be complemented by a systematic scientific evaluation of the safety of people occupationally exposed in agriculture or public health. The essential scientific information for this assessment of occupational safety of pesticides is the detailed toxicology of the active ingredients and the acute toxicity and irritancy of the formulation, together with information on exposure during use. The uses of pesticides which are permissible, on the basis of occupational safety, are defined by relating information on occupational exposure to the level of exposure which toxicity studies show to be potentially harmful. The relationship between these two sets of information is the intrinsic margin of safety between the measurable occupational exposure to a pesticide and the level of exposure which, if exceeded, potentially may cause harm. The toxicity data must, therefore, identify dose levels which are demonstrably toxic, and lower dose levels at which there is no observable toxicity, i.e. a no-effect-level by the relevant exposure route.

Margins of Safety

There are many theoretically possible permutations of toxicity of different types. Likewise there are many ways in which pesticides might practically be used (or misused). Hence, there is no simple equation or check list which describes the minimum acceptable margin of safety. In some instances the acceptable safety margin for occupational exposure may be relatively small; usually it is very large. The nature of the toxic hazard, and the size of the dose which would be harmful, strongly influence decisions on permitted uses and, hence, safety margins. A reversible biochemical response, such as slight cholinesterase inhibition, might require a margin of safety which is only small provided the benefit outweighs the risk, as in the case of malaria vector control by insecticide spraying described in Chapter 2. Crop uses of a cholinesterase inhibiting pesticide would involve substantially larger safety margins. If the toxic hazard was a reversible minor change, seen in several species, then a safety margin of at least a factor of ten would be expected for the worst likely occupational exposure and absorption. A larger margin would be needed if there was a lesion which was a cumulative response, slow to reverse, and was evident after repeated dermal exposure in animal toxicity tests. Irreversible or delayed toxicity requires a substantial margin of safety. For example, a teratogenic hazard or a weak carcinogenic effect in long term rodent studies by a non-genotoxic compound given at high dose levels would be acceptable for specified uses provided occupational exposure evidently was several hundredfold less than a potentially hazardous level. Major differences in toxicity between mammalian species in laboratory studies generally would indicate the need for an increase in the safety margin unless there was evidence from metabolic or pharmacokinetic studies that man resembled the less susceptible species.

With a formulation which is irritant to the skin or eye in the non-diluted form, or one which causes sensitisation, use is only possible if protective gloves and eye protection can be worn when handling the concentrate. Toxicity studies will show whether the spray dilution also is irritant. Active ingredients which easily penetrate the skin, and have serious acute or cumulative effects at moderate doses, require handling precautions which strictly control contact with human skin. If there is substantial penetration of the skin by an active ingredient there will be only a minimal difference in toxicity by the dermal and the oral (or parenteral) routes in laboratory animals. If the rate of penetration by the active ingredients is particularly rapid in dermal toxicity tests there will be early onset of symptoms and the dermal route toxicity will not increase very much when the contact time is extended from, for example, one to six or even 24 hours. Dermal toxicity studies are conducted on both the active ingredient and the formulation, and so both absorption and toxic effects of skin contamination are investigated qualitatively and quantitatively without resort to experiments on humans.

Occupational Exposure Studies

There are less clearcut instances when the scientific evaluation of the toxicology leads to the judgement that a particular use of a certain pesticide probably has an adequate margin of safety for those occupationally exposed, but perhaps some doubt remains. In such situations information on occupational exposure is an essential part of the safety evaluation. This information can identify the benefits of the safety precautions, including any protective clothing which may be specified on the label for the pesticide. Hence, exposure studies or calculations have a place in safety evaluations for pesticides.

The safety evaluation of pesticides, including safety precautions, labelling and the many aspects mentioned in this and earlier chapters, is soundly based in several separate scientific disciplines. A theoretical relationship can be seen between many of the factors described in this book as affecting occupational exposure and, hence, safety of pesticides in use. However, there are gaps in our understanding, including the prediction of absorbed dose from information on skin contamination as measured by the pad or suit-analysis techniques, although this gap has been circumvented by the tried and tested use of dermal toxicity data on the active ingredient and the formulation.

Bioavailability of pesticide on the skin

Not all the contamination measured on the uncovered skin or under the work clothing is biologically available. Except in the case of spillages which saturate the clothing, pads worn under the clothing in exposure studies probably consistently over-estimate the actual contamination of the adjacent skin. Unlike skin, these pads are highly absorbent and they are raised above the body surface. Drops of spray or formulation landing on the clothing are absorbed into the clothing (and the pad) and dry in situ. Subsequent transfer to the skin will depend, in part, upon the type of fabric of the clothing, and the rate of evaporation which in turn depends upon the temperature and the nature of the vehicle or solvent. Similarly, small drops sporadically landing on the unclothed skin will dry, perhaps unnoticed.

Skin acts as a depot and even after decontamination procedures (e.g. washing) some absorption continues. Exfoliation of the superficial cells as part of the normal skin turn-over will deplete the depot, but by how much?

Exposure to dilute aqueous sprays deposited over a period of time as small drops is not, in terms of safety, equivalent to the same amount of contamination arriving in an instant by splashing or direct contact, and remaining there available for absorption for the rest of the time period. Both are measured in exposure studies as potential skin exposure or actual skin contamination and are not distinguished by chemical analysis methods. Techniques which visualise the contamination in situ on the clothing by, for example, coloured or fluorescent dyes, may be able to separate contamination by direct contact, which gives a discrete area of contamination, from spray drop contamination which appears as isolated foci.

Guy and Maibach (1984) concluded that the data available for interpretation and analysis are minimal and that extrapolations and generalisations must be very cautiously advanced. For occupational exposure, there may sometimes be an interest in the daily and weekly absorbed dose. If this is calculated from exposure studies the result commonly will over-estimate the real situation and the margin of safety will appear smaller than it actually is. However, only rarely is precise knowledge of the absorbed dose (and hence the precise margin of safety) needed since generally exposure is far below levels which may be hazardous. Absorption studies are not a prerequisite for safety evaluations.

In summary, it is unwise to assume that a steady state of absorption from skin contamination prevails, or that a mean rate of absorption calculated from permeability or octanol/water partition coefficients is an accurate description of events under field conditions. Extrapolation between exposure studies and absorption tends to over-estimate the absorbed dose and so if there is concern that the margin of safety is too small then some form of measurement of absorption is needed.

CONCLUSIONS

The record of safe use of pesticides in those countries with reliable statistics is a good one. Registered pesticides, even those of high toxicity, are safely used. That is not, of course, ground for complacency. There is always the need for better information on accidents and their causes and for early warnings of bad use practices. Nowhere is this need more evident than in some developing countries, about which assertions of frequent neglect of safe use practices and allegations of widespread injury are often made.

It is, of course, essential that everything possible be done to educate the users of pesticides to become skilled and careful workers. All concerned, government, industry and workers themselves, have vital roles to play in reducing accidents. Unnecessary exposure by misuse is pointless and every effort must be made to eliminate it. The need is for unremitting effort in all the areas that were described as affecting exposure. That effort is not glamorous or particularly newsworthy but it is the best means to control occupational exposure to pesticides.

The occupational hazards which may result from working with pesticides are assessed routinely, by government authorities, industry and others, on the basis of the toxicology and the likely exposure.

Pesticides constitute an important beneficial input to agriculture and public health. Moreover, they can be and are used safely. Unfortunately, this is not the view of the more extravagant critics whose only proposals seem to consist of restrictions and bans. If they have their way, it is the nutrition and health of the very people and countries that they think they are protecting that will suffer.

REFERENCES

Advisory Committee on Pesticides (1980) Further review of the safety for use in the UK of the herbicide 2,4,5-T. Ministry of Agriculture Fisheries and Food, London.

Anon. (1978) DBCP, Chlordecone and the risk benefit equation. Lancet, No. 8080, 79-80.

Balarajan, R. and McDowall, M. (1983) Congenital malformations and agriculatural workers. Lancet, No. 8333, 1112-3.

Blair, A., Grauman, D.J., Lubin, J.H. and Fraumeni, J.F. (1983) Lung cancer and other causes of death among licensed pesticide applicators. J. Nat. Lanc. Inst., 71, 31-36.

Caufield, C. (1984) Pesticides: exporting death. New Scientist, No. 1417, 15-17.

Coffee, R.A. (1981) Electrodynamic crop spraying. Outlook on Agriculture, 10, 350-356.

Espir, M.L.E., Hall, J.W., Shirreffs, J.G. and Stevens, D.L. (1970) Impotence in farm workers using toxic chemicals. Br. Med. J., Feb. 14th, 423-5.

FAO (1982) Pesticide Residues in Food, U.N. Food and Agriculture Organisation, Evaluations 1981.

FAO (1983a) Guidelines on Good Labelling Practice for Pesticides. FAO Plant Protection Bulletin, 31, 71-89.

FAO (1983b) Guidelines for the packaging and storage of pesticides. FAO Plant Protection Bulletin, 31, 63-69.

Finley, E.L. et al. (1974) Efficacy of home laundering in removal of DDT, methyl parathion and toxaphene residues from contaminated fabrics. Bull. Environmental Contamination and Toxicology, 12, 268-274.

Friends of the Earth (1984) Pesticides: the case of an industry out of control. Section 6 Risks to Users. London.

Golding, J. and Sladden, T. (1983) Congenital malformations and agriculatural workers. Lancet, No. 8338, 1393.

Gusman, S. and Irwin, F. (1980) Chemical Hazard Warnings - Labelling for Effective Communication. The Conservation Foundation, Washington.

Guy, R.H. and Maibach, H.I. (1984) Correction factors for determining body exposure from forearm percutaneous absorption data. J. Applied Toxicology, 4, 26-28.

Hardell, L. and Sandstrom, A. (1979) Case controlled study: Soft tissue sarcomas and exposure to phenoxy acetic acid or chlorophenols. Br. J. Cancer, 39, 711-717.

Hayes, W.J. (1982) Pesticides studied in man. Williams and Wilkins, London and Baltimore.

International Organisation of Consumer Unions/Friends of the Earth, Malaysia (1982); see also the IOCU press release, Holland, Feb. 1984.

IARC (1979, 1983) Monographs: Evaluation of Carcinogenic Risk of Chemicals to Man, 20, 45-349 and 30, 61-344 respectively.

Jacobs, W.W. (1984) Closed systems for mixing and loading. In Determination and Assessment of Pesticide Exposure. Studies in Environmental Science, 24, 155-169.

Jellinek, S.D. (1984) In Handbook of Chemical Industry Labelling pp 281-299, Editors C.J. O'Connor and S.I. Lirtzman, Noyes Data Corporation, New Jersey.

Kilgore, W.W. and Akesson, N.B. (1980) Minimising occupational exposure to pesticides: populations at exposure risk. Residue Reviews, 75, 21-31.

Lirtzman, S.I. (1984) In Handbook of Chemical Industry Labelling pp 5-40, Editors C.J. O'Connor and S.I. Lirtzman, Noyes Data Corporation, New Jersey.

Murphy, S.D. (1980) Pesticides, In Toxicology: The Basic Science of Poisons, pp 357-408, Editors J. Doull, C.D., Klaassen, and M.O. Amdur, Macmillan, London.

Pesticide Container Design Panel (1984) Draft - Guidelines for the design and handling of containers for liquid agrochemicals prepared under the aegis of the Pesticide Registration Department; Ministry of Agriculture Fisheries and Food.

Reeves, J.D., Driggers, D.A. and Kiley, V.A. (1981) Household insecticide associated aplastic anaemia and acute leukemia in children. Lancet, No. 8241, 300-1.

Staiff, D.C., Davis, J.E. and Stevens, E.R. (1982) Evaluation of various clothing materials for protection and acceptability during application of pesticides. Arch. Environmental Contamination and Toxicology, 11, 391-398.

Torkelson, T.R., Sadek, S.E. and Rowe, V.K. (1961) Toxicologic investigations of 1,2-dibromo-3-chloropropane. Tox. Appl. Pharm., 3, 545-559.

Vettorazzi, G. (1979) International Regulatory Aspects for Pesticide Chemicals, 1, Toxicity Profiles, CRC Press, Florida.

WHO (1981) Toxicology of pesticides, In Health Aspects of Chemical Safety, Interim document, 9.

WHO Data Sheets on Pesticides (1975-1984) Nos. 1-58, issued as VBC/DS documents.

Whorton, D., Krauss, R.M., Marshall, S. and Milby, T.H. (1977) Infertility in male pesticide workers. Lancet No. 8051, 1259-61.

APPENDIX 1

WORKER EXPOSURE TO PESTICIDES DURING USE

S.J. Crome

INTRODUCTION

This Appendix summarises data on dermal and respiratory exposure of workers to any type of pesticide in commercial use in agriculture or public health. It presents the detailed figures from which the summary tables in Chapter 3 are compiled. Not included in the compilation are accounts not published in English or studies in which exposure was assessed by indirect methods, e.g. by measuring cholinesterase inhibition or blood or urine levels of pesticide metabolites. Papers which omit important details of the method also were excluded. These self-imposed limitations were necessary in view of the diversity of published literature and the need for detailed review of each publication.

The main data tables present the total dermal and inhalational exposure as measured in the original reference with the necessary back-up data to put the figures in context and allow comparisons. The general and specific notes provide further information on the methods used.

The literature was obtained from a collection of papers covering the period since 1955 supplemented by on-line search of the TOXLINE, MEDLINE, BIOSIS, CAB ONLINE, EXCERPTA MEDICA and SCIENCE CITATION INDEX data bases using a range of keywords.

FORMAT

There are six tables which summarise the pesticide, the type and concentration of formulation, the rate of application, the type of use and the exposure data. Also cited are the original reference, and notes explaining particular features of the study.

The type of formulation is described as emulsifiable concentrate (EC), wettable powder (WP), granules, dusts or aerosol. "Spray" or "liquid concentrate" or "liquid spray" is given when the reference does not detail the type of formulation. For references before the mid 1960s it is unlikely that many sprays were suspension concentrates (SC). "Low volume" spray refers to liquid formulations which are applied undiluted or with relatively little dilution.

To permit easy application of the data to different types of formulation and use, the exposure values in the original references have been converted into mg of formulation or ml of spray per person per hour of actual or potential contamination. Values for mg of active ingredient are quoted wherever the paper does not give the necessary information to convert. Respiratory exposure is expressed in mg formulation/person/h, ml spray/person/h or mg active ingredient/person/h for the same reasons.

GENERAL NOTES TO THE TABLES

These notes give, in detail, the conditions under which the exposure figures were measured and the method of measurement. They are designed to cover method of measurement, type of clothing worn and type of operations contributing to the exposure and for skin exposure to distinguish estimated actual skin contamination from measured potential skin exposure.

A) Dermal exposure figures are mean and range (minimum - maximum in parenthesis), range only, below the limits of analytical detection (ND), or the average (AV) from the original papers. For Table A exposure is as mg formulation/person/h wherever possible and mg active ingredient/person/h when the necessary data for translation from active ingredient to formulation is not available. Comparisons based upon exposure per hour during tank filling or loading the formulation may not be the most appropriate when the individuals involved work quickly and take very little time for each filling operation. The number of filling operations in a work period is always an important factor. However a days work may, in fact, involve about an hour in total doing the filling operation. For Tables B-F exposure is as ml spray/person/h wherever a liquid was applied. This would be spray dilution, or neat formulation if no dilution occurs. If a powder was applied the figures are mg powder/person/h. If translation from mg active ingredient was not possible mg active ingredient person/h figures are given instead as indicated in the table heading.

B) Respiratory exposure figures are mean and range (minimum - maximum in parenthesis), range only, maximum only (MAX), below the limits of analytical detection (ND), or the average (AV) from the original papers. Units are as described above for dermal exposure.

C) Dermal exposure figures are the estimated actual skin contamination. Values were calculated by extrapolating from compound impinging on pads to that which would impinge on unprotected skin as calculated by the original authors. The original authors assumed that normal clothing including overalls, boiler suits etc., provides absolute protection to areas covered.

D) Dermal exposure figures are the estimated actual skin contamination. In these cases values were calculated by extrapolating from compound impinging on pads to that which would impinge on unprotected skin plus that which would penetrate normal clothing including overalls, boiler suits etc., as calculated by the original authors.

E) Exact details of the calculation methods used are not given by the original authors. Hence the distinction between actual or potential skin exposure is not certain.

F) Dermal exposure figures are for the estimated actual skin contamination when protective clothing is used. This may have been specially impermeable clothing e.g. rubber aprons, or merely "more protective than normal" clothing, e.g. long-sleeved instead of short-sleeved shirts.

G) For details of dermal exposure assessment see specific notes.

H) Respiratory route exposure figures are calculated from the amount of chemical impinging on a respirator filter.

I) Respiratory exposure figures are calculated by measuring the concentration of compound in the air in the operator's breathing zone and assuming normal breathing. If both H and I appear in the General Notes column a combination of both methods has been used.

L) Operators wore no specific protective clothing.

M) Details of the clothing worn are not given in the original reference.

N) Operators wore full or partial protective clothing; see specific notes and "F" above.

P) Figures include filling equipment, as well as actual spraying. Such values are therefore not included in the summary table.

Q) Figures are for spraying (or dusting) only, not filling or mixing.

R) Figures are for filling tanks only.

S) Details of operations contributing to exposure are not given.

T) Operations contributing to exposure did not include spraying or mixing/filling.

*) Extra data given under the indicated specific note.

†) Dermal exposure values so marked are for potential skin exposure, as distinct from notes C, D and F above. Relevant supplementary information is given in Specific Notes 8, 12, 13, 16, 17, 22, 33, 35, 39, 43, 44 and 46 for potential skin exposure, and in Specific Notes 2, 11, 19, 21 and 26 for estimated actual skin contamination.

TABLES

In the Tables there are separate section for the types of formulation, the activities involved and specific features of the studies as appropriate, as follows:

Exposure to products while loading or diluting.............A

Exposure while applying using vehicle mounted equipment - upward directed spray or dust..............................B

Exposure while applying using tractor mounted equipment - downward directed spray or dust............................C

Exposure from knapsack sprayers and other hand-held appliances - outdoor use...................................D

Exposure from knapsack sprayers and other hand-held appliances - indoor use....................................E

Exposure during application from aircraft..................F

TABLE A EXPOSURE TO PRODUCT WHILE LOADING OR DILUTING

COMPOUND	TYPE OF FORMULATION	TYPE OF USE	DERMAL EXPOSURE (A) mg form/person/h	RESPIRATORY EXPOSURE (B) mg form/person/h	REF	GENERAL NOTES	SPECIFIC NOTES
LIQUID FORMULATIONS							
Phospholan	250* EC	Mixing 1.5 L pesticide to 300 L water	110.4		1	DLR	1, 5
Ethion	46.5 EC	Diluting to 0.06% for air blast spraying	615 (20-2955)	-	47	CLR	5
Ethion	46.5 EC	Diluting to 0.09% for air blast spraying	75.5 (70.5-80.4)	-	47	CLR	5
Diallate	45 EC	Filling tractor tanks before sugar beet spraying - manual system	AV = 3258	AV = 0.0051	33	CILR	5
Diallate	45 EC	- closed system	AV = 0.482	0.0002-0.00056	33	CILR	5
Diallate	45 EC	- manual system, gloves worn	AV = 15	AV = 0.0051	33	CINR	5
EPN	36 EC	Diluting chemical and loading aircraft	0.28*	0.0058	39	GILR	5, 25
EPN	48 EC	Diluting chemical and loading aircraft	3.83	0.0048	39	CILR	5
EPN	36 EC	Diluting chemical and loading aircraft	1.14*	0.0036	39	GILR	5, 25

TABLE A (Continued) EXPOSURE TO PRODUCT WHILE LOADING OR DILUTING

COMPOUND	TYPE OF FORMULATION	TYPE OF USE	DERMAL EXPOSURE (A) mg form/person/h	RESPIRATORY EXPOSURE (B) mg form/person/h	REF	GENERAL NOTES	SPECIFIC NOTES
Dimefox	50% liquid formulation	Diluting to 0.5% spray dilution in 50 gallon tanks – air at worker's position, head height	–	AV = 0.12 MAX = 0.48	19	INR	18, 5, 8
		– air around tanks at head height	–	AV = 0.16 MAX = 0.40	19	I	18, 5
		Diluting to 1% spray dilution in 50 gallon tanks – air at worker's position, head height	–	AV = 0.08 MAX = 0.20	19	INR	18, 5, 8
		– air around tanks at head height	–	AV = 0.12 MAX = 0.60	19	I	18, 5
Carbaryl	48% liquid concentrate	Mixing and loading helicopter	1.0	–	50	C(F)NR	2, 5
Carbaryl	48% liquid concentrate	Mixing and loading helicopter	3.1	–	50	C(F)NR	2, 5
Paraquat	14.5% concentrate	Diluting to 0.63% for knapsack spraying	0.7–19.3	0.00083	49	CILR	5
Lead arsenate	32.8% liquid formulation	Diluting to 0.1% for air blast spraying	180 ± 64	0.015	66	CHLR	3

Lead arsenate	32.8% liquid formulation	Diluting to 0.3% for air blast spraying	591 ± 143	0.015	66	CHLR	3
2,4-D	48% concentrate	Mixing for, and loading, helicopter without protective clothing - closed system	8.1 (0.38-17.2)	MAX = 0.05	32	CILR	5, 21
2,4-D	48% concentrate	Mixing for, and loading, helicopter wearing protective clothing - closed system	2.25 (1.00-4.08)	ND	32	FINR	5, 21
2,4,5-T	48% concentrate	supervisor/mixer filling backpack sprayer	1.5	ND	34	CILR	5, 23
2,4,5-T	48% concentrate	Mixing for tractor mist blower	64.0	0.0075	34	CILR	
2,4,5-T	48% concentrate	Supervisor for tractor mist blower	7.9	0.004	34	CILQ	
2,4,5-T	48% concentrate	Mixer loading helicopter	7.1 (3.64-10.5)	ND	34	CILR	5, 23
2,4,5-T	48% concentrate	Supervisor of loading helicopter	0*-2.10	ND	34	CILQ	5, 23
Carbaryl	40 SC*	Mixing and loading tractor mounted boom sprayer	95 (53.3-174.5)	-	50	CLR	5, 29
Carbaryl	40 SC*	As above	10.3 (1.0-13.5)	-	50	C(F)NR	5, 2 29
Carbaryl	40 SC*	Mixing and loading for helicopter	19 (3.3-45.3)	-	50	C(F)NR	5, 2 29

TABLE A (Continued) EXPOSURE TO PRODUCT WHILE LOADING OR DILUTING

COMPOUND	TYPE OF FORMULATION	TYPE OF USE	DERMAL EXPOSURE (A) mg form/person/h	RESPIRATORY EXPOSURE (B) mg form/person/h	REF	GENERAL NOTES	SPECIFIC NOTES
Mevinphos	47.1 EC	Mixing and filling for boom spraying – closed system	–	0.00011–0.03	56	IR	5, 37
Azinphos-methyl	25 WP	Tank filling of air-blast machine	211.6 (40.4–316)	5.08 (0.4–12.5)(H) 4.92 (0.48–11.0)(I)	25	C HILR	9, 5
Parathion	25 WP	Concentration in air, storage warehouse when opened in morning = 0.003 ppm					
		Man standing in warehouse without disturbing bags would be exposed to:	–	0.14	14	I	5
The figures below are for potential skin exposure, not actual skin contamination (see General Note †)							
2,4,D	48% concentrate	Mixing/loading tractor drawn boom sprayer	†612.6 (72.0–1369.8)	–	61	GR	44
2,4,D	48% concentrate	Mixing/loading tractor mounted boom sprayer	†1465.2 (570–2915.4)	–	61	GR	44
2,4,D	48% concentrate	Mixing/loading tractor mounted CDA applicator	†730.8 (216.6–1417.8)	–	61	GR	44

2,4,D	48% at concentrate 0.7% dilution	Loading pre-diluted chemical into knapsack sprayers	†72.6 (46.2-285.6)	-	61	GR	44
Chlorobenzilate	40 EC	Diluting to 0.108% for air blast spraying	†20.8 ± 6.5	0.02 ± 0.005	65	GHNR	22, 5
Ethion	46.5 EC	Diluting to 0.06% for air blast spraying	†4955 (163-23832)	0.011 (0.004-0.024)	47	GHLR	12, 5
Ethion	46.5 EC	Diluting to 0.09% for air blast spraying	†609 (570-649)	0.017	47	GHLR	12
Paraquat	14.5% concentrate	Diluting to 0.63% for knapsack spraying	†93.3	0.00083	49	GILR	13, 5
Prochloraz	45 EC	Mixing and loading for tractor trailed and self propelled boom sprayer	†2982 (203-5249)	≤1.8 (ND-3.5)	68	IR	46
<u>POWDERS OR DUSTS</u>							
Methomyl	24.1% Powder	Mixing and filling for boom spraying - closed system	-	ND - 2.76	56	IR	5, 36 37, 18
Acephate	75% Water soluble powder	Mixing and filling for boom spraying	-	ND - 0.22	56	IR	5, 37 18
Benomyl	50 WP	Mixing and loading aircraft	624 (94.8-1464)	1.92 (0.72-5.52)	40	CHLR	5
Terbutryne	80 WP	Mixing for boom spraying	4960	9.79	44	CILR	5
Terbutryne	80 WP	Mixing for boom spraying	10.6	-	44	FNR	2, 5

TABLE A (Continued) EXPOSURE TO PRODUCT WHILE LOADING OR DILUTING

COMPOUND	TYPE OF FORMULATION	TYPE OF USE	DERMAL EXPOSURE (A) mg form/person/h	RESPIRATORY EXPOSURE (B) mg form/person/h	REF	GENERAL NOTES	SPECIFIC NOTES
Carbaryl	80 WP	Mixing and loading tractor mounted boom sprayer	250.1 (95.3 - 436.4)	-	50	CLR	5
Carbaryl	80 WP	As above	50.6 (44.3 - 64.5)	-	50	C(F)NR	2, 5
Carbaryl	80 WP	Mixing and loading for helicopter	56.6 (19.6 - 87.3)	-	50	CLR	5
Carbaryl	50 WP	Mixing and loading for hand-held lance	123.6	-	50	CLR	5
Carbaryl	50 WP	Mixing and loading for hand-held lance	20.9	-	50	CLR	5
TEPP	1% dust	Loading aeroplane	7300 (4300-13600)	15 (3-43)	6	CHILR	5
The figures from the data below cannot be translated from mg a.i./person/h.							
Methidathion	EC	Diluting and loading for air blast sprayer	0.003	0.0005	45	FINR	2, 5
Methidathion	EC	Diluting and loading for boom sprayer	0.16	0.0005	45	FINR	2, 5

Parathion	Liquid Concentrate and WP	Loading or mixing for hand-held sprayer	-	0.38 (ND-2.76)	16	IR	7, 18 5
Parathion	Liquid Concentrate and WP	Loading or mixing for air-blast machine	-	0.26 (Trace-1.22)	16	IR	7, 18 5
Parathion	?	Ground crews servicing aeroplanes spraying various crops	0.04-5.0	0.047-0.176	35	EINR	2, 5
Quintozene	Powder undiluted	Filling tanks on tractor for use in forest nurseries	-	0.8	11	IR	18, 5
Quintozene	Powder undiluted	Filling manual applicator for use in forest nurseries	-	1.0	11	IR	18, 5

TABLE B EXPOSURE WHILE APPLYING USING VEHICLE MOUNTED EQUIPMENT - UPWARD DIRECTED SPRAY OR DUST

COMPOUND	TYPE OF FORMULATION AND DILUTION	RATE OF APPLICATION L spray/ha	TYPE OF USE	DERMAL EXPOSURE (A) ml spray/ person/h	RESPIRATORY EXPOSURE (B) ml spray/ person/h	REF	GENERAL NOTES	SPECIFIC NOTES
SPRAYING LIQUID								
Ethion	46.5% EC 0.06%	4732	Air-blast spraying of citrus	59.5 (3-228)	-	47	CLQ	
Ethion	46.5% EC 0.09%	4732	Air blast spraying of citrus	1084 (148-2021)	-	47	CLQ	5
Chlorobenzilate	40 EC 0.108%	2338	Air blast spraying of fruit orchards - operator	35 $\pm$ 6	0.012 $\pm$ 0.002	65	FHLQ	3
Methidathion	EC 0.06%	18710	Air blast spraying in orchards	1.78	0.012	45	FINQ	2, 5
Dimethoate	? EC 0.09% spray	16840-18710	Air blast spraying of fruit orchards:					
			open topped tractor	-	0.021	52	IQ	18, 32
			cab tractor with windows open	-	0.026	52	IQ	18, 32
			cab tractor with windows shut	-	0.004	52	IQ	18, 32
Azinphos-methyl	Spray 0.05%	6732	Air blast machine spraying fruit orchards - operator	54 (2.2-292)	0.08 (0.04-0.16)	6	CHI LQ	5

DDT	Spray 0.09%	9963	Air blast machine spraying fruit orchards – operator	60 (3.6–436)	0.1 (0.02–0.30)	6	CHI LQ	5
Dieldrin	Spray 0.02–0.03%	7473–14010	Power air blast spraying of fruit orchards – operator	62 (25.2–124.4)	0.12 (0.08–0.16)	6	CHI LQ	5
DNOC	Spray 0.02–0.04%	3082–11770	Thinning apple blossom by air blast spraying – operator	75 (9.7–437)	0.17 (0.13–0.27)	6	CHI LQ	5
Malathion	Spray 0.04–0.08%	4204–11208	Air blast spraying of fruit orchards – operator	50 (9.8–98)	0.18 (0.03–0.4)	6	CHI LQ	5
Parathion	Spray 0.05%	4484–6726	Air blast machine spraying of citrus groves – operator	36 (2.6–76)	0.06 (0.02–0.14)	6	CHI LQ	5
Carbaryl	Liquid spray 0.045–0.06%	?	Air blast spraying of fruit orchards – operator	111 (3.2–400)	0.17 (0.02–2.04)	23	CHL Q	
Lead arsenate	32.8% liquid formulation 0.1%	4700	Air blast spraying in orchards	55 ± 9	0.002	66	CHLQ	5
Lead arsenate	32.8% liquid formulation 0.3%	2067	Air blast spraying in orchards	32 ± 4	0.002	66	CHLQ	5
Na-DNOC	19% Aqueous slurry 0.02–0.08	4677–11220	Spraying apples for blossom thinning	AV = 126.4	AV = 0.8	17	CHL S	
Carbaryl	50 WP 0.06–0.96%	?	Air blast spraying of apples – tractor driver	4.96 (3.5–5.94)	0.056 (0.047–0.104)	3	CHL P	5
Parathion	15 WP 0.048–0.48%	?	Air blast spraying of apples – tractor driver	0.92 (0.27–2.23)	0.012 (0.0038–0.019)	3	CHL P	5

TABLE B (Continued) EXPOSURE WHILE APPLYING USING VEHICLE MOUNTED EQUIPMENT - UPWARD DIRECTED SPRAY OR DUST

COMPOUND	TYPE OF FORMULATION AND DILUTION	RATE OF APPLICATION L spray/ha	TYPE OF USE	DERMAL EXPOSURE (A) ml spray/ person/h	RESPIRATORY EXPOSURE (B) ml spray/ person/h	REF	GENERAL NOTES	SPECIFIC NOTES
Parathion	25 WP 0.12%	17957	Spraying olives, details not given. Air sample from near spray operators.	-	0.069	14	IQ M	6, 5
Parathion	25 WP 0.12%	15900	Spraying olives, details not given. Air sample from near spray operators.	-	0.175	14	IQ M	6, 5
Parathion	25 WP 0.18%	935	Spraying walnut trees with tractor drawn sprayer - driver	-	0.039	14	IQ M.	6, 5
Parathion	25 WP 0.144%	17770	Spraying olives, details not given. Air sample from near spray operators.	-	0.14-0.20	14	IQ M	6, 5
Parathion	Conventional parathion formulation 0.03%	3742-9354	Air blast spraying of orchards	AV = 64.7	AV = 0.07	4	CH LS	
Azinphos-methyl	25 WP 0.06-0.72%	?	Air blast spraying of fruit orchards - operator	3.2 (0.28-17.9)	0.067 (0.018-0.41) (H) 0.077 (0.051-0.28) (I)	25	CHI LQ	9, 5

Parathion	? WP 0.044% spray	11230- 18710	Air blast spraying of fruit orchards:					
			open topped tractor	-	0.041	52	IQ	18, 32
			cab tractor with windows open	-	0.023	52	IQ	18, 32
			cab tractor with windows shut	-	0.007	52	IQ	18, 32
Azinphos-methyl	50 WP 0.18-0.25%	562-787	Spraying orchards with air blast equipment	0.007*	0.012	36	FINQ	2, 14
Azinphos-methyl	50 WP 0.18-0.25%	562-787	Spraying orchards with air blast equipment	0.002-0.007*	0.012	36	FINQ	2, 14
Azinphos-methyl	50 WP 0.18-0.25%	562-787	Spraying orchards with air blast equipment	0.001-0.004*	0.012	36	FINQ	2, 14
Parathion	? 0.24-0.36%	140-935	Air blast spraying of orchards - operator	13.0	0.026	4	CHLQ	27
Methidathion	EC 0.06%	18710	Upwards directed spraying in orchards	0.78	0.018	45	FINQ	2, 5
Parathion	48 EC 0.045%	16840- 18710	Oscillating boom spraying of fruit orchards	-	0.004-0.104	52	IQ	18, 32
Demeton	Spray 0.03%	7473	Tractor pulled high pressure power hand gun sprayer in nursery - driver	6.3 (3.3-8.3)	0.03 (0.03-0.10)	6	CHI LQ	5
Parathion	Spray 0.05%	4484-6726	Tractor pulled portable power hand gun sprayer in citrus groves - tractor driver	24 (11-50)	0.06 (0.02-0.12)	6	CHI LQ	5

TABLE B (Continued) EXPOSURE WHILE APPLYING USING VEHICLE MOUNTED EQUIPMENT - UPWARD DIRECTED SPRAY OR DUST

COMPOUND	TYPE OF FORMULATION AND DILUTION	RATE OF APPLICATION L spray/ha	TYPE OF USE	DERMAL EXPOSURE (A) ml spray/ person/h	RESPIRATORY EXPOSURE (B) ml spray/ person/h	REF	GENERAL NOTES	SPECIFIC NOTES
Parathion	25 WP 0.192%	1871	Spraying oranges with tractor drawn sprayer - driver	-	0.52	14	IQM	6, 5
Carbaryl	80 WP 0.225%	?	Spraying trees with power sprayer	56.2 (1.37-178.3)	0.044 (ND-0.088)	48	DILQ	
Carbaryl	80 WP 0.225%	?	Spraying trees with power sprayer	26.4	0.044	48	DILP	
2,4-D + 2,4,5-T	? emulsion 2%	?	Spraying in forests with tractor driven equipment - operator	-	0.025-0.05	30	IQ	18
2,4,5-T	48% concentrate 2.4% spray	93.5	Spraying in forests using tractor mounted mist blower - driver	1.02 (0.78-1.25)	0.00021 (0.000019-0.00024	34	CILQ	5
2,4,5-T	? 4.8% spray	46.8	Spraying in forests using tractor mounted mist blower - driver	0.016-0.073	-	31	CLP	5

Malathion	ULV 2.5–5.0%	0.75–1.5	Cold fogging against mosquitoes with jeep mounted aerosol generator - operator (spraying and mixing)	0.16–0.44	0.028–0.063	15	FINP	2, 18
Malathion	ULV 2.5–5.0%	0.75–1.5	as above - second operator (spraying only)	0.031–0.074	0.028–0.063	15	FINQ	2, 18
Malathion	ULV 2.5–5.0%	0.75–1.5	as above - first operator (spraying and mixing)	0.87–2.03	-	15	CLP	19
Malathion	ULV 2.5–5.0%	0.75–1.5	as above - second operator (spraying only)	0.063–0.14	-	15	CLQ	19
Chlorthion	ULV 2.5–5.0%	0.75–1.5	as above - first operator (spraying and mixing)	0.093–0.15	0.018–0.037	15	FINP	2, 18
Chlorthion	ULV 2.5–5.0%	0.75–1.5	as above - second operator (spraying only)	0.013–0.063	0.018–0.037	15	FINQ	2, 18
Chlorthion	ULV 2.5–5.0%	0.75–1.5	as above - first operator (spraying and mixing)	0.18–0.32	-	15	CLP	19
Chlorthion	ULV 2.5–5.0%	0.75–1.5	as above - second operator (spraying only)	0.05–0.15	-	15	CLQ	19
Malathion	ULV 2.5–5.0%	0.75–1.5	Standing in field observing drift of spray	0.038–0.95	-	15	FNT	2
Malathion	ULV 2.5–5.0%	0.75–1.5	Standing in field observing drift of spray	0.06–0.18	-	15	CLT	19
Chlorthion	ULV 2.5–5.0%	0.75–1.5	Standing in field observing drift of spray	0.07–0.012	-	15	FNT	2

TABLE B (Continued) EXPOSURE WHILE APPLYING USING VEHICLE MOUNTED EQUIPMENT - UPWARD DIRECTED SPRAY OR DUST

COMPOUND	TYPE OF FORMULATION AND DILUTION	RATE OF APPLICATION L spray/ha	TYPE OF USE	DERMAL EXPOSURE (A) ml spray/ person/h	RESPIRATORY EXPOSURE (B) ml spray/ person/h	REF	GENERAL NOTES	SPECIFIC NOTES
Chlorthion	ULV 2.5-5.0%	0.75-1.5	Standing in field observing drift of spray	0.02-0.043	-	15	CLT	19
Chlorthion	5% aerosol	?	Operating machine for mosquito control	0.14 (0.038-0.24)	0.0048 (0.0016-0.01)	6	CHILQ	5, 28
Malathion	2.5-5% aerosol	?	Operating machine for mosquito control	0.77 (0.099 - 1.41)	0.0019 (0.0011-0.0024)	6	CHILQ	5, 28
The figures below are for potential skin exposure, not actual skin contamination (see General Note †)								
Ethion	46.5% EC 0.06%	4732	Air-blast spraying of citrus	†480 (125-1838)	0.008 (0.005-0.015)	47	GHLQ	12, 5
Ethion	46.5% EC 0.09%	4732	Air blast spraying of citrus	†8744 (1189-16300)	0.004 (0.001-0.009)	47	GHLQ	12, 5
DUST APPLICATION				g dust/ person/h	g dust/ person/h			
DDT + BHC	5% + 1% dust	50 kg/ha	Horse-drawn dusting in forest for moth control:			13	FINP	2
			DDT	4.3	0.098			
			BHC	7.0	0.31			

Parathion	2% dust	112 kg/hg	Sample taken in dusted area during treatment of orange trees	-	0.058	14	IT	18, 5

The figures from the data below cannot be translated from mg a.i./person/h.

Parathion	Liquid spray ?	0.22-1.1 kg a.i./ha	Air blast spraying of fruit orchards - operator	AV = 77.7	AV = 0.16	16	CHL P	7, 5
Parathion	?	?	Air blast spraying of orchards, tractor driver	1.61-93.9	AV = 0.019	2	EH MQ	
Na-DNOC	Liquid concentrate	1.46 kg/ha	Spraying apples for blossom thinning	AV = 24.4	0.03	5	CHL S	5
Quintozene	Powder undiluted	10-50 kg/ha	Dusting in forest nurseries with tractor drawn equipment - driver	-	0.65	11	IQ	18, 5
			- assistant	-	0.25	11	IQ	2, 18
Azinphos-methyl	?	?	Air blast spraying of fruit orchards - operator	AV = 9.94	0.102 (0.045-1.5)	26	CHL S	
Carbaryl	?	?	Air blast spraying of fruit orchards - operator	AV = 24.9	0.48 (0.01-1.08)	26	CHL S	
MCPA	? powder 1-2%	?	Spraying in forests (farmers)	-	0.005 (ND-0.115)	67	IP	38
Dichlorprop	? liquid 1-2%	?	Spraying in forests (farmers)	-	0.009 (ND-0.038)	67	IP	38
Mecoprop	? liquid 1-2%	?	Spraying in forests (farmers)	-	ND (ND-0.033)	67	IP	38

TABLE B (Continued) EXPOSURE WHILE APPLYING USING VEHICLE MOUNTED EQUIPMENT - UPWARD DIRECTED SPRAY OR DUST

COMPOUND	TYPE OF FORMULATION AND DILUTION	RATE OF APPLICATION L spray/ha	TYPE OF USE	DERMAL EXPOSURE (A) mg a.i./ person/h	RESPIRATORY EXPOSURE (B) mg a.i./ person/h	REF	GENERAL NOTES	SPECIFIC NOTES
MCPA	? powder 1-2%	?	Spraying in forests (professional sprayers)	-	0.019 (0.005-0.03)	67	IP	38
Dichlorprop	? liquid 1-2%	?	Spraying in forests (professional sprayers)	-	0.005 (0.001-0.041)	67	IP	38
Mecoprop	? liquid 1-2%	?	Spraying in forests (professional sprayers)	-	0.002-0.018	67	IP	38

TABLE C EXPOSURE WHILE APPLYING USING TRACTOR MOUNTED EQUIPMENT – DOWNWARD DIRECTED SPRAY OR DUST

COMPOUND	TYPE OF FORMULATION AND DILUTION	RATE OF APPLICATION L spray/ha	TYPE OF USE	DERMAL EXPOSURE (A) ml spray/ person/h	RESPIRATORY EXPOSURE (B) ml spray/ person/h	REF	GENERAL NOTES	SPECIFIC NOTES
SPRAYING LIQUIDS								
Endrin	20 EC 0.05%	2806-3274	Air blast or hand held sprayers on orchard cover crops.	AV = 5.2	AV = 0.02	7	CHL S	4
Endrin	Spray 0.05%	2690	Air blast or boom sprayers on orchard cover crops – operator.	5 (2.6-12.2)	0.02 (0.002-0.04)	6	CHI LQ	5
Diallate	45 EC 2.22% spray	140	Spraying sugar beet using a spray/harrow apparatus with – closed cab tractor	AV = 0.003	AV = 0.00035	33	CILQ	5
			– open cab tractor	–	AV = 0.00029	33	I	5
Diallate	45 EC 2.22% spray	140	Spraying sugar beet using conventional boom sprayer – closed cab tractor	AV = 0.010	AV = 0.00011	33	CILQ	5
			– open cab tractor	–	AV = 0.00009	33	I	5
EPN	60 EC 2.15% spray	51.5	Spraying cotton	0.012-0.074	0.00007-0.0004	39	CILS	5

TABLE C (Continued) EXPOSURE WHILE APPLYING USING TRACTOR MOUNTED EQUIPMENT - DOWNWARD DIRECTED SPRAY OR DUST

COMPOUND	TYPE OF FORMULATION AND DILUTION	RATE OF APPLICATION L spray/ha	TYPE OF USE	DERMAL EXPOSURE (A) ml spray/ person/h	RESPIRATORY EXPOSURE (B) ml spray/ person/h	REF	GENERAL NOTES	SPECIFIC NOTES
Methidathion	EC 2.4% spray	187.1	Spraying alfalfa with boom sprayers	0.058	0.00033	43	FINQ	2
Nitrofen	25 EC 0.31-1.37%	168-561	Boom spraying of ground crops	0.0043	0.00022	57	FINP	2, 38, 39
Nitrofen	25 EC 0.31-1.37%	168-561	Boom spraying of ground crops	0.0045	0.00039	57	FINP	2, 38, 40
Nitrofen	25 EC 0.31-1.37%	168-561	Boom spraying of ground crops	0.35	0.0022	57	GINP	38, 39, 40
Endrin	20% emulsion 0.15-0.3%	?	Downwards directed air-blast spraying of ground crops - tractor driver.	0.066 (0-0.13)	ND	3	CHMQ	5
Dinoseb	36% "liquid concentrate" 0.09%	281-561	Tractor mounted spray boom for weed control, tractor driver.	AV = 98.6	AV = 0.13	5	CHL P	5
Dinoseb	36% "liquid concentrate" 0.09%	281-561	Open cab truck with spray boom attached near front bumper, driver.	AV = 17.3	-	5	CLQ	5
Dinoseb	36% "liquid concentrate"	281-561	Tractor with front spray boom, driver.	AV = 2.5	-	5	CLQ	5

Dinoseb	36% "liquid concentrate" 0.09%	281-561	Tractor with rear spray boom, driver.	AV = 4.6	-	5	CLQ	5
Dicamba + 2,4-D	Both 40% liquid formulations 0.01% + 0.013%	44.5	Boom spray of pasture - tractor driver dicamba + isomer 2,4-D	- -	0.0039-0.0107 0.0052-0.0096	53	IQ	18
Paraquat	24% liquid concentrate 0.5%	935	Boom spraying of ground crops in orchards	0.08 (0.02-0.68)	0.0002 (ND-0.0004)	58	CHLQ	5
Parathion	Spray 0.09%	623	Tractor mounted boom ground sprayer in row crops - operator.	5.2 (2.4-15.0)	AV = 0.011	6	CHI LQ	5
2,4,5-T	? 4.8% spray	46.8	Spraying rice levees using tractor mounted sprayer - driver	0.062	-	31	CLP	5
Nitrofen	50 WP 0.31-1.37%	165-561	Boom spraying of ground crops	0.0077	0.005	57	FINP	2, 37, 40
Nitrofen	50 WP 0.31-1.37%	165-561	Boom spraying of ground crops	0.46	0.49	57	GINP	38, 39, 40
DDT	75% ai formulation 1% spray	?	Spraying conifer seedlings	-	0.0005-0.01*	46	IQ	15
DDT	75% ai formulation 2% dip	?	Dipping conifer seedlings into solution	-	0.00025-0.019	46	IQ	

TABLE C (Continued) EXPOSURE WHILE APPLYING USING TRACTOR MOUNTED EQUIPMENT - DOWNWARD DIRECTED SPRAY OR DUST

COMPOUND	TYPE OF FORMULATION AND DILUTION	RATE OF APPLICATION L spray/ha	TYPE OF USE	DERMAL EXPOSURE (A) ml spray/ person/h	RESPIRATORY EXPOSURE (B) ml spray/ person/h	REF	GENERAL NOTES	SPECIFIC NOTES
Captan	50 WP 0.071% spray	1403	Spraying strawberries using tractor mounted boom	18.8	0.65	63	CHLP	
The figures below are for potential skin exposure, not actual skin contamination (see General Note †)								
2,4-D	48% concentrate 0.7% spray	200	Boom spraying-tractor drawn sprayer	†40.3 (10.1-104.0)	ND	61	GIQ	43
2,4-D	48% concentrate 0.7% spray	200	Boom spraying-tractor mounted sprayer	†42.7 (14.6-87.5)	ND	61	GIQ	43
2,4-D	48% concentrate 3.2% spray	45	CDA application-tractor mounted sprayer	†96.4 (35.3-145.7)	ND	61	GIQ	43
Nitrofen	25 EC 0.31-1.37%	168-561	Boom spraying of ground crops	†0.064	0.014	57	GINP	37, 38, 39
Phenmedipham	11.4% EC undiluted	120	Spraying sugar beet with tractor mounted boom sprayer - driver	†0.16-0.34	0.00024-0.001	9	GHNQ	16

Phenmedipham	11.4% EC undiluted	240	Spraying sugar beet with tractor mounted boom sprayer - driver	†0.058-0.168	0.00003-0.00008	9	GHNQ	16
Prochloraz	45 EC 0.24% spray	200	Spraying wheat with conventional boom sprayer	†0.9 (0.3-2.2)	≤0.07 (ND-0.2)	68	IQ	46
DUST APPLICATION				g dust/ person/h	g dust/ person/h			
Malathion	4% dust	4.483 kg/hg	Power duster applying to pole beans	0.58 (0.43-0.80)	0.018 (0.0055-0.0308)	6	CHILQ	5
Parathion	1% dust	0.34-0.44 kg/hg	Tractor mounted ground duster in row crops	0.88 (0.140-1.70)	0.016 (0.003-0.041)	6	CHILQ	5
Endrin	1% dust	?	Dusting potatoes	1.87	0.0041	7	CHLS	
Parathion	1% dust	28 kg/ha	Tractor driver during dusting	-	11	14	IT	18, 5
Captan	5% dust	-	Planting treated seed potatoes with tractor driven planter - driver - observer	0.007 ± 0.004	0.0007 ± 0.0004 -	38	CHLQ	42
GRANULE APPLICATION								
Thionazin	10% granule	-	Planting granules in furrow - machine operator	0.0050-0.0075	AV = 0.000014 MAX = 0.000048	20	CILQ	18
Disulfoton	5% granule	-	Applying to rows of growing crops - tractor driver	0.010-0.015	AV = 0.000054-0.0018 MAX = 0.00034-0.00046	20	CILQ	18, 20

TABLE C (Continued) EXPOSURE WHILE APPLYING USING TRACTOR MOUNTED EQUIPMENT - DOWNWARD DIRECTED SPRAY OR DUST

COMPOUND	TYPE OF FORMULATION AND DILUTION	RATE OF APPLICATION L spray/ha	TYPE OF USE	DERMAL EXPOSURE (A) g dust/ person/h	RESPIRATORY EXPOSURE (B) g dust/ person/h	REF	GENERAL NOTES	SPECIFIC NOTES
Phorate	10% granule	-	Applying to rows of growing crops - tractor driver	0.005-0.0075	AV = 0.000076-0.00031 MAX = 0.00009-0.00046	20	CILQ	18, 20
The figures from the data below cannot be translated from mg a.i./person/h.								
Carbaryl	40 SC*	3.36 kg/ha	Applying to potatoes - tractor driver	2.8 (0.3-8.5)	-	50	CLQ	5, 29
Mevinphos	? liquid formulation	?	Boom spraying - tractor driver	-	0.0008-0.005	56	IQ	5, 18
Terbutryne	80 WP ?	2.69 kg ai/ha	Pre-emergence application to sorghum by boom sprayer	12.3	0.29	44	CILQ	5
Terbutryne	80 WP ?	2.69 kg ai/ha	Pre-emergence application to sorghum by boom sprayer	1.6	-	44	FNQ	2, 5
Carbaryl	80 WP	3.36 kg/ha	Applying to peas and potatoes - tractor driver	107 (0.03-4.65)	-	50	CLQ	5
Methomyl	powder	?	Boom spraying - tractor driver	-	0.0038	56	IQ	5, 18
Acephate	water soluble powder	?	Boom spraying - tractor driver	-	0.0008-0.003	56	IQ	5, 18

TABLE D EXPOSURE FROM KNAPSACK SPRAYERS AND OTHER HAND - HELD APPLIANCES - OUTDOOR USE

COMPOUND	TYPE OF FORMULATION AND DILUTION	RATE OF APPLICATION L spray/ha	TYPE OF USE	DERMAL EXPOSURE (A) ml spray/ person/h	RESPIRATORY EXPOSURE (B) ml spray/ person/h	REF	GENERAL NOTES	SPECIFIC NOTES
SPRAYING LIQUIDS								
Demeton	Spray 0.03%	7480	Using tractor pulled high pressure power hand gun sprayer in nursery.	10.3 (5.3-19.3)	0.03 (0.03-0.1)	6	CHI LQ	5
Dieldrin	Spray 0.03%	9340	Hand gun spraying of orchards with portable machine - operator.	50.4 (11.3-98.3)	0.1 (0.07-0.13)	6	CHI LQ	5
DNOC	Spray 0.02-0.04% 0.03	3082-11770	Thinning apple blossom by power hand gun spraying - operator.	183.7 (23.3-300.7)	0.43 (0.07-1.4)	6	CHI LQ	5
Endrin	Spray 0.05%	2690	High pressure hand gun spraying of orchard cover - operator.	6 (3-14.2)	0.02 (0.002-0.06)	6	CHI LQ	5
Malathion	Spray 0.03-0.08 0.055	4204-14943	High pressure hand gun spraying of orchards	121.8 (15.3-352.7)	0.1 (0.018-0.45)	6	CHI LQ	5
Parathion	Spray 0.05%	4484-6726	Tractor-pulled portable power hand gun sprayer in citrus - operating sprayer from a tower.	22 (2.0-5.6)	0.06 (0.008-0.10)	6	CHI LQ	5

TABLE D (Continued) EXPOSURE FROM KNAPSACK SPRAYERS AND OTHER HAND - HELD APPLIANCES - OUTDOOR USE

COMPOUND	TYPE OF FORMULATION AND DILUTION	RATE OF APPLICATION L spray/ha	TYPE OF USE	DERMAL EXPOSURE (A) ml spray/ person/h	RESPIRATORY EXPOSURE (B) ml spray/ person/h	REF	GENERAL NOTES	SPECIFIC NOTES
Parathion	Spray 0.05%	4484-6726	Tractor-pulled portable power hand gun sprayer in citrus - operating sprayer from ground.	94 (40-226)	0.18 (0.04-0.38)	6	CHI LQ	5
2,4-D	28 EC or 35 EC 1%	?	Spraying ground weeds using gun fed by vehicle mounted tanks	-	0.0009-0.002	37	IQ	
Picloram	9.3 EC 1%	?	Spraying ground weeds using gun fed by vehicle mounted tanks	-	0.0002-0.0003	37	IQ	
Dichlorprop	28 EC 1%	?	Spraying ground weeds using gun fed by vehicle mounted tanks	-	0.001-0.002	37	IQ	
2,4-D	28 EC or 35 SC 6%	?	Backpack mist blower to spray weeds	-	0.001	37	IQ	
Picloram	9.3 EC 6%	?	Backpack mist blower to spray weeds	-	0.0003	37	IQ	
Dichlorprop	28 EC 6%	?	Backpack mist blower to spray weeds	-	0.002	37	IQ	

Phospholan	"250* EC" 1%	357	Solo Low Volume Knapsack Sprayer to spray cotton.	0.86–1.17	–	1	DLQ	1, 5
Phospholan	"250* EC" 0.5%	714	John Bean High Volume Spray motor to spray cotton – tractor mounted reservoir but hand held nozzle.	0.58–1.22	–	1	DLQ	1, 5
Parathion	? 0.31%	?	Spraying tomatoes using knapsack sprayer	AV = 2.9	0.09 (0.008–0.4)(H) 0.02–0.06 (I)	27	CHI LP	18, 10
Dimefox	50% liquid formulation 0.5%	153–166	Spraying hops with hand held sprayer fed from tank.	–	0.0008 (0.006 –0.012)	19	IMQ	18, 5
Dimethoate	32% liquid spray 0.13%	357	Spraying aubergines, alfalfa, akra, tomatoes, cucumbers, melons using a knapsack sprayer.	0.63 (0.03–1.6)	0.0018 (0.001–0.0038)	28	DHL Q	11
Fenthion	? 0.06%	?	Spraying against mosquitoes with portable power sprayer	6.0 (0.17–19.0)	<0.027 (<0.0011–0.15)	59	CHLQ	
Fenthion	? 0.06%	?	Spraying against mosquitos with portable hand pressure sprayer	6.0 (0.17–10.5)	<0.033 (<0.0017–0.11)	59	CHLQ	
Paraquat	14.5% concentrate 0.63% spray	135–169	Spraying rubber plantations using knapsack sprayers	0.35 (0.016–1.97	0.00008	49	CILQ	5
2,4,5-T	48% concentrate 1.9% spray	93.5	Spraying in forests – backpack sprayer operator	1.55 (0.25–3.12)	0.002 (0.0004 –0.008)	34	CILQ	5

TABLE D (Continued) EXPOSURE FROM KNAPSACK SPRAYERS AND OTHER HAND - HELD APPLIANCES - OUTDOOR USE

COMPOUND	TYPE OF FORMULATION AND DILUTION	RATE OF APPLICATIÓN L spray/ha	TYPE OF USE	DERMAL EXPOSURE (A) ml spray/ person/h	RESPIRATORY EXPOSURE (B) ml spray/ person/h	REF	GENERAL NOTES	SPECIFIC NOTES
Diazinon	25 EC 0.047%	1.22 L/m^2	Spraying lawns with compressed air sprayers			64		
			- hands only	11.7 ± 8.5				
			- total dermal except hands	0.045 ± 0.049				
			- total dermal	11.7 ± 8.5	0.004 ± 0.005		CHLP	
Diazinon	25 EC 0.063%	Wet to run off	Spraying shrubs with compressed air sprayers			64		
			- hands only	10.9 ± 9.1				
			- total dermal except hands	0.19 ± 0.23				
			- total dermal	11.2 ± 9.6	0.005 ± 0.004		CHLP	
Diazinon	25 EC 0.062%	1.22 L/m^2	Spraying lawns using hosepipe-powered sprayer			64		
			- hands only	40.5 ± 59.9				
			- total dermal except hands	0.87 ± 0.07				
			- total dermal	41.4 ± 59.9	0.012 ± 0.016		CHLP	
Benomyl	50 WP 0.2% spray	?	Spraying ornamental bushes in gardens	1	0.005	40	CHLP	5
Benomyl	50 WP 0.2% spray	?	Spraying vegatable gardens	2.1	0.0057	40	CHLP	5

DDT	75 WP 5%	40 ml/m^2	Outdoor house spraying against mosquitoes with hand held sprayer.	AV = 4.86	AV = 0.0022	18	CNL S	5
2,4,5-T	? 2.4% spray	93.5	Spraying in forests - operator	0.079-1.80	-	31	CLP	5
A pyrethroid	2.5% liquid undiluted	2	Spraying cotton	0.59 (0.096-1.69)	0.002-0.004	12	CILQ	18
The figures below are for potential skin exposure, not actual skin contamination (see General Note †)								
2,4-D	48% concentrate 0.7% spray	200	Spraying grass using hand-held boom with 4 nozzles	†141.8 (98.0-191.3)	ND	61	GIQ	43
2,4-D	48% concentrate 0.7% spray	?	Spraying in forestry using hand-held single nozzle lance	†84.0 (67.7-110.9)	0.005	61	GIQ	43
Paraquat	14.5% concentrate 0.63% spray	135-169	Spraying rubber plantations using knapsack sprayers	†10.5 (1.92-27.0)	0.00008	49	GILQ	13, 5
GRANULE/DUST APPLICATION								
Fenthion	1% granule	?	Hand application to control mosquitos	1.23 (0.28-5.88) g form/p/h	0.0088 (0.0004-0.0186) g form/p/h	59	CHLQ	
The figures from the data below cannot be translated from mg a.i./person/h.								
Parathion	Liquid spray ?	0.22-1.1 kg a.i./ha	Spraying orchards using hand held sprayer supplied from tank.	AV = 56	AV = 0.19	16	CHL Q	7, 5

TABLE D (Continued) EXPOSURE FROM KNAPSACK SPRAYERS AND OTHER HAND - HELD APPLIANCES - OUTDOOR USE

COMPOUND	TYPE OF FORMULATION AND DILUTION	RATE OF APPLICATION L spray/ha	TYPE OF USE	DERMAL EXPOSURE (A) mg a.i./ person/h	RESPIRATORY EXPOSURE (B) mg a.i./ person/h	REF	GENERAL NOTES	SPECIFIC NOTES
The figures from the data below cannot be translated from mg a.i./person/h.								
Carbaryl	? SC* 1% spray	1 L/m^3	Spraying tree boles with a stirrup pump against bark beetles	†62.7 mg/person/ application	-	55	GNQ	2, 34, 35
Paraquat	0.44% pressurised hand spray*	spot appln	To weeds in yards and gardens	0.066 (0.002-0.13)	<0.0002 (ND-<0.0002)	58	CHLQ	5, 41
Carbaryl	50 WP	4.48 kg/ha	Lance spraying of apple trees	22.8	-	50	CLQ	5
Carbaryl	50 WP	2.24 kg/ha	Lance spraying of apple trees	22.8	-	50	CLQ	5
Quintozene	powder undiluted	10-50 kg/ha	Manual application to forest nurseries - operator	-	1.75	11	IQ	18, 5
DDT + BHC	15% DDT + 50% BHC ?	?	Spraying full grown trees from ground using portable sprayer.	-	AV = 20.6	13	INS	3

TABLE E EXPOSURE FROM KNAPSACK SPRAYERS AND OTHER HAND - HELD APPLIANCES - INDOOR USE

COMPOUND	TYPE OF FORMULATION AND DILUTION	RATE OF APPLICATION L spray/ha	TYPE OF USE	DERMAL EXPOSURE (A) ml spray/ person/h	RESPIRATORY EXPOSURE (B) ml spray/ person/h	REF	GENERAL NOTES	SPECIFIC NOTES
Benomyl	50 WP 0.2% spray	?	Spraying beans, cucumber and cotton in greenhouses	0.32	0.0017	40	CHLP	5
Bis-ethyl-mercury phosphate	technical solid 0.00625%	5470 ml/m^2	Treating 170 m^3 glasshouse tomatoes with tank-fed hand-held sprayer.	-	AV = 0.048 MAX = 0.128	22	INQ	2, 18
			Treating 17m^3 glasshouse tomatoes using 2 gallon watering can.	-	0.004-0.016	22	INQ	2, 18
Chlorpyrifos (a) + Dichlorvos (b) mixture	a) 23.5% EC 1.8% dilution b) 23% EC 1.13% dilution	? ?	Spraying houses to kill cockroaches, ants etc.	-	a) 0.0008 (0.0001-0.0037) b) 0.0004 (0.00009-0.0012)	29	IMS	18
Chlorpyrifos	EC 0.5% spray	16.4 ml/m^3	Spraying dormitories against pests	-	0.000012	42	IQ	18
Acephate	EC 1% spray	18.5 ml/m^3	Spraying dormitories against pests	-	0.00007	42	IQ	18

TABLE E (Continued) EXPOSURE FROM KNAPSACK SPRAYERS AND OTHER HAND - HELD APPLIANCES - INDOOR USE

COMPOUND	TYPE OF FORMULATION AND DILUTION	RATE OF APPLICATION L spray/ha	TYPE OF USE	DERMAL EXPOSURE (A) ml spray/ person/h	RESPIRATORY EXPOSURE (B) ml spray/ person/h	REF	GENERAL NOTES	SPECIFIC NOTES
Diazinon	EC 1% spray	18.0 ml/m^3	Spraying dormitories against pests	-	0.00008	42	IQ	18
Fenitrothion	EC 1% spray	21.9 ml/m^3	Spraying dormitories against pests	-	0.00016	42	IQ	18
Propoxur	EC 1.1% spray	18.5 ml/m^3	Spraying dormitories against pests	-	0.0007	42	IQ	18
DDT	75 WP 5%	40 ml/m^2	Indoor house spraying against mosquitoes with a hand held sprayer.	AV = 35.1	AV = 0.142	18	CHL S	5
Dieldrin	? WP 0.55%	90 ml/m^2	Spraying in houses against mosquitos using a stirrup pump	AV = 0.33	-	24	FNS	2
Bendiocarb	80 WP 1%	40 ml/m^2	Spraying walls and ceilings for mosquito control	1.99 (0.19-3.19)	-	8	DNQ	2
Bendiocarb	WP 0.5% spray	19 ml/m^2	Spraying dormitories against pests	-	0.0008	42	IQ	18
Methomyl	? 0.25%	118 ml/m^2	Wand sprayer attached to large tank spraying in	-	0.176-1.91	51	IQ	18

Chlorpyrifos	2% slow release formulation	?	Painted on surfaces to kill cockroaches in houses	0.34	0.0003	41	CILS	26
Chlorpyrifos	2% slow release formulation spray applied	?	Sprayed to kill cockroaches in houses	0.64	0.0018	41	CILS	26
Permethrin	? WP 1-2%	–	Dipping conifer seedlings into solution	–	0.00067	60	IQ	18
			Packing dipped seedlings	–	0.0004–0.0029	60	IQ	18
			Planting dipped seedlings	–	0.00007 (0.00003–0.0002)	60	IT	18
Diazinon	2% dust	?	Treating attics against roaches, silverfish and fleas – operator	–	0.19 (0.02–1.05) mg form/p/h	29	IP	18
			– supervisor	–	0.03	29	IT	18
Carbaryl	5% dust	1260 mg/m^3	Dusting dormitories against pests	–	0.014 mg form/p/h	42	IQ	18
Thiram	? liquid flowable form. ?	*	Seed treatment using specialised machinery	<0.5–3.7 mg a.i./h	<0.5–0.75 mg a.i./h	62	DHL T	45
Carboxin	? liquid flowable form. ?	*	Seed treatment using specialised machinery	<0.5–8.62 mg a.i./h	<0.5–0.88 mg a.i./h	62	DHL T	45
Maneb + lindane	? dust	*	Seed treatment using specialised machinery	54.8–81.2 mg a.i./h	0.36–0.54 mg a.i./h	62	DHL T	45

TABLE E (Continued) EXPOSURE FROM KNAPSACK SPRAYERS AND OTHER HAND - HELD APPLIANCES - INDOOR USE

COMPOUND	TYPE OF FORMULATION AND DILUTION	RATE OF APPLICATION L spray/ha	TYPE OF USE	DERMAL EXPOSURE (A) ml spray/ person/h	RESPIRATORY EXPOSURE (B) ml spray/ person/h	REF	GENERAL NOTES	SPECIFIC NOTES
The figures below are for potential skin exposure, not actual skin contamination (see General Note †)								
Bendiocarb	25% ULV	0.18 ml/m^2	Sprayed on floors to kill cockroaches in					
			- warehouse	†16.2	0.008	54	GIQ	18, 33
			- barn	†2.2	0.005	54	GIQ	18, 33
			- food processing room	†13.3	0.0014	54	GIQ	18, 33
Bendiocarb	80 WP 3%	40 ml/m^2	Spraying lower walls and floors for flea control	†a 1.49-8.2 mg a.i./p/ trial †b 10.5-14.4 mg a.i./p/ trial	-	10	GNQ	17

TABLE F EXPOSURE DURING APPLICATION FROM AIRCRAFT

COMPOUND	TYPE OF FORMULATION AND DILUTION	RATE OF APPLICATION L spray/ha	TYPE OF USE	DERMAL EXPOSURE (A) ml spray/ person/h	RESPIRATORY EXPOSURE (B) ml spray/ person/h	REF	GENERAL NOTES	SPECIFIC NOTES
EPN	36 EC 2.4% spray	28	Spraying cotton from aircraft - pilot	0.0073	0.00004	39	CILQ	5
EPN	36 EC 3% spray	28	Spraying cotton from aircraft - pilot	0.0037	0.00004	39	CILQ	5
EPN	36 EC 2.4-3% spray	28	Flagging for above spraying	1.04	0.0029	39	CILT	5
EPN	48 EC 4% spray	28	Spraying cotton from aircraft - pilot	0.016	0.00004	39	CILQ	5
EPN	48 EC 4% spray	28	Flagging for above spraying	0.22	0.036	39	CILT	5
EPN	36 EC 3% spray	28	Spraying cotton from aircraft - pilot	0.011	0.00005	39	CILQ	5
EPN	36 EC 3% spray	28	Flagging for above spraying	0.125	0.00006	39	CILT	5
2,4-D	48% concentrate 5% spray	93.5	Application to forests from helicopters - pilots	0.00088 (0-0.0014)	ND*	32	CILQ	5, 21
		93.5	- mechanics	0.042 (0.0034 -0.11)	ND*	32	CILT	5, 21

TABLE F (Continued) EXPOSURE DURING APPLICATION FROM AIRCRAFT

COMPOUND	TYPE OF FORMULATION AND DILUTION	RATE OF APPLICATION L spray/ha	TYPE OF USE	DERMAL EXPOSURE (A) ml spray/ person/h	RESPIRATORY EXPOSURE (B) ml spray/ person/h	REF	GENERAL NOTES	SPECIFIC NOTES
2,4-D	48% concentrate 5% spray	93.5	- supervisors	0.0003 (0-0.0009)	ND*	32	CILT	5, 21
		93.5	- observers standing outside spray area	0.00012 (0.000074)	ND*	32	CILT	5, 21
Parathion	9% spray	?	Flagging for airplane application in orchards	0.93 (0.11-3.33)	0.00022 (0.00003-0.00089)	6	CHILT	5
2,4-D	48% concentrate 5% spray	93.5	Application to forest from helicopters					
			- pilots	0.00006 (0-0.0013)	ND*	32	CINP	5, 21
			- mechanics	0.011 (0.0001 -0.031)	ND*	32	CINT	5, 21
			- supervisors	0.002 (0.0008 -0.005)	ND*	32	CINT	5, 21
			- observers standing outside spray area	0.0022 (0-0.0033)	ND*	32	CINT	5, 21

2,4,5-T	48% concentrate 4.8% spray	46.7	Helicopter application to forests - pilot	ND*-0.096	ND*	34	CILQ	5, 23
2,4,5-T	48% concentrate 4.8% spray	46.7	Helicopter application to forests - flagmen	ND*	0-0.0009	34	CILT	5, 23
2,4,5-T	?	46.8	Spraying against weeds in rice. Average for pilot, mixer, flaggers and supervisor	0.0027-0.010	-	31	CLP	5
The figures below are for potential skin exposure, no actual skin contamination (see General Note †)								
Fenitrothion	30% ULV undiluted	<5	Piloting plane spraying forests	†0.00011-0.00012	<0.00001*	21	IQ	39
DUST APPLICATION				g dust/person/h	g dust/person/h			
DDT	35% dust	19.58 kg/ha	Flagging for airplane dusting of orchards	1.2 (1.13-1.48)	0.0006	6	CHILT	5
Parathion	2% dust	1.12 kg/ha	Piloting airplane dusting orchards	0.65 (0.42-0.95)	0.001 (0.001-0.002)	6	CHILQ	5
TEPP	1% dust	0.56 kg/ha	Piloting plane dusting orchards	2.40 (1.0-5.30)	0.017 (0.002-0.047)	6	CHILQ	5
TEPP	1% dust	0.56 kg/ha	Flagging for above dusting	0.60 (1.60-2.10)	0.007 (0.003-0.012)	6	CHILT	5
Parathion	2% dust	112 kg/ha	Sample taken in dusted area during helicopter dusting of olive trees	-	0.0135	14	IT	18, 5

TABLE F (Continued) EXPOSURE DURING APPLICATION FROM AIRCRAFT

COMPOUND	TYPE OF FORMULATION AND DILUTION	RATE OF APPLICATION L spray/ha	TYPE OF USE	DERMAL EXPOSURE (A) g dust/ person/h	RESPIRATORY EXPOSURE (B) g dust/ person/h	REF	GENERAL NOTES	SPECIFIC NOTES
GRANULE APPLICATION								
Disulfoton	5% granule	-	Application to row crops					
			- pilot	0.01-0.015	MAX = 0.000028	20	CILQ	18, 5, 20
			- groundmarkers	0.0434	MAX = 0.00042	20	CILT	18, 5, 20
The figures for the data below cannot be translated from mg a.i./person/h.								
				mg a.i./ person/h	mg a.i./ person/h			
Carbaryl	48% liquid concentrate	2.24 kg/ha	Helicopter with boom, spraying corn - pilot	366.5*	-	50	CLQ	5, 30
			- flagger	605.7	-	50	CLT	5
Carbaryl	48% liquid concentrate	1.12 kg/ha	Helicopter with boom, spraying corn - pilot	3.4	-	50	CLQ	5
			- flagger	407.9	-	50	CLT	5
Carbaryl	40 SC*	3.36 kg/ha	Helicopter with boom, spraying corn - pilot	26.5 (5.1-74.6)	-	50	CLQ	5, 29
			- flagger	384.7 (218.0 530.0)	-	50	CLT	5

Carbaryl	80 WP	2.24 kg/ha	Helicopter with boom, spraying corn - pilot	7.4 (2.4–15.9)	-	50	CLQ	5
			- flagger	176.9 (115.1–272.1)	-	50	CLT	5, 31
Parathion	?	?	Aeroplane application to various crops - pilot	0.005–0.2	0.094 (0.009–0.62)	35	EILQ	5, 24

SPECIFIC NOTES

These specific notes supplement figures in the Tables and the General Notes.

1) * "250 EC" quoted, assumed to be 25% active ingredient in the formulation (250 mg/ml).

2) Protective clothing was in use when the data was produced. The amount of clothing varies from reference to reference.

3) The quoted respiratory exposure is calculated from air concentrations and men should have worn masks therefore the figure is for potential exposure only.

4) No significant difference between air blast and hand-gun spraying found when data treated separately.

5) Figures from this reference appear in more than one of Tables A to F.

6) Respiratory exposure figures calculated assuming 1 ppm = 11.9 mg/m^3 for parathion, and breathing rate 0.50 m^3/hour.

7) Comparison of methods "H" and "I" (see General Notes) for calculating respiratory exposure while spraying gives

	Air blast spraying	Hand-held sprayer
respirator pads	0.16 mg/man/h	0.19 mg/man/h
air concentration	0.04 mg/man/h	0.11 mg/man/h

Note approximate 1.5-4 fold difference.

Respiratory exposure values for mixing (Table A) are means of 27 samples of ordinary wettable powders (WP), antidusting wettable powder (AWP) and liquid concentrates (LC).

8) Respirator and protective clothing were worn during mixing, but figures are potential exposure as if no respirator worn.

9) The two respiratory figures are - top calculated by respirator technique (See note H) - lower calculated by air intake technique (See note I) but assuming

respiratory rate of 0.444 m^3/hr.

10) Clothing worn even less that usual; shorts replaced long trousers and feet left bare.

11) Clothing worn was North African "Aregie", a knee length shirt worn with only a head cloth and tennis shoes. Values for exposure through this clothing are included in the figure in the table.

12) Dermal exposure figures are total amount of chemical found on bare skin plus that on top of clothing.

13) Dermal exposures are amounts of chemical which would cover entire body as if no clothes were worn.

14) *Method of assessing dermal exposure used here is unusual. Figure reflects amount of chemical which would penetrate protective clothing (when worn) plus normal clothing underneath. It does not include amount which would be found on uncovered areas. This explains very low figures.

15) * Exposures given based on air concentrations 7 days after spraying. Reference states - "Field application gives negligible airborne concentrations".

16) Body surface exposure calculated by assessing how much chemical landed on full boiler suit with hood. No allowance made for chemical penetrating boiler suit or landing on unprotected skin.

17) Dermal exposure in mg/person/hour could not be calculated as lengths of trials not given. Exposure measured by measuring chemical on boiler suit with hood. Exposure of uncovered areas not included. Exposure calculated assuming surface area of 1.8 m^2/person.

    Figures (a) were while wearing wellington boots, figures (b) were without such boots.

18) Respiratory exposure figures extrapolated from air concentrations using breathing rate of 0.5 m^3/h.

19) Dermal figures quoted measured by measuring chemical on protective clothing and then extrapolated to normal clothing.

20) Respiratory exposure ranges established by summarising individual data points obtained using formulations containing various carriers. Attapulgite resulted in greatest inhalational exposure and pumice in greatest contact exposure.

21) Figures given for when protective clothing in use reflect amount of chemical which would still reach bare skin in spite of specialised clothing.

    * ND = below minimum level of detection (approximately 0.00015 mg/person/hour based on average exposure times, body weights and volumes of air sampled. A breathing rate of 1.74 m^3/hour is assumed).

    In this study an automatic mixing/filling system was in use. This system broke down on one occasion and tanks were filled by hand. This resulted in the only respiratory exposure above the minimum level of detection shown in Table A. Dermal exposure was not noticeably altered.

22) Rubber aprons and cloth lined rubber gloves were in use. Pads were placed outside the clothing but under the aprons.

23) * For respiratory and dermal exposure ND = below minimum level of detection.

24) Details of procedures to estimate dermal exposure very sketchy. Pads were used. Influence of clothing on exposure is not established.

25) * Dermal exposure figure is exposure of hands only.

26) Figure for dermal exposure calculated from given data using exposed body areas only. Total exposure (i.e. exposed areas plus that inside clothes) as given in original data was 64.9 mg/person/hour when formulation was sprayed.

    No mixing involved, but paper does not make it clear if filling of equipment included in measurements.

27) An observation of the authors was that low-volume sprays contaminating clothing dry much more quickly. This could allow resulting dusty residue to work its way through to the skin unnoticed while the considerable wetting of a high volume spray would be noticed and clothing changed.

28) Aerosol generation was probably by cold-fogging, but not stated in paper.

29) * Formulation a 40% ai suspension in molasses.

30) * Abnormally high value caused by equipment failure (other values measured under similar conditions generally <10 mg/person/h) showing danger from such events quite clearly.

31) Flagger in this test stood at edge of sprayed field rather than in it (compare with other values from Ref. 50), hence lower exposure.

32) This paper contains much interesting information on effects of tractor cabs on dermal and respiratory exposure during application of parathion and dimethoate to orchards. Dermal data unfortunately not in a form from which whole body exposure can be easily obtained, hence no figures in main tables.

33) Dermal exposure assessed by measuring total chemical on whole of disposable boiler suit. Thus exposure takes no account of protection from any clothing.

34) Liquid formulation in molasses (may have been 40%).

35) Values in mg/person/hour cannot be calculated from available data. Values given are whole body exposure on top of protective clothing as estimated by dermal pads; they bear no resemblance to chemical which would impinge on bare skin.

36) ND = non detected; 16 out of 18 replicates were ND. The 0.666 mg/p/hr was caused when a water soluble bag of powdered formulation broke during loading.

37) Closed transfer systems were used during mixing and filling.

38) Respiratory exposures calculated from air concentrations at a breathing rate of 1.25 m^3/hour.

39) Dermal exposure is chemical found outside protective clothing over the whole body, not just on "normally uncovered" areas.

40) For filling using an open transfer system.

41) * Application was probably by commercial standard aerosol can.

42) Figure in table relates to planting of previously treated potatoes.

43) Potential skin exposure values are for chemical on unprotected skin, plus chemical on top of normal working clothes.

44) Potential skin exposure values are for chemical on unprotected skin, not including head and neck, plus chemical on top of normal working clothes.

The reference gives data as mg formulation per filling; this has been converted to mg formulation/hour assuming each filling of the tank took 10 minutes.

45) *Data are for using machinary to treat seed at 2,200-58,000 kg seed/hour. All measurable dermal exposure was to the hands. 0.5 mg a.i./h was the limit of detection.

46) Potential skin exposure was assessed by measuring the amount of chemical landing on a boiler suit with a hood. Respiratory exposure figures were extrapolated from measured air concentrations on the basis of breathing 1.7 m^3/h.

REFERENCES

1) Soliman S.A. et al., Occupational Effect of Phospholan Insecticide on Spraymen during Field Exposure, J. Environ; Sci. and Health, 14, 27-37, 1979.

2) Durham W.F., Wolfe H.R., An additional Note Regarding Measurement of the Exposure of Workers to Pesticides, Bull. W.H.O., 29, 279-281, 1963.

3) Jegier Z., Health Hazards in Insecticide Spraying of Crops, Arch. Environ. Health, 8, 670-674, 1964.

4) Wolfe H.R., Armstrong J.F., Durham W.F., Pesticide Exposure from Concentrate Spraying, Arch. Environ. Health, 13, 340-344, 1966.

5) Wolfe H.R., Durham W.F., Batchelor G.S., Health Hazards of some Dinitro Compounds, Arch. Environ. Health, 3, 468-475, 1961.

6) Wolfe H.R. et al., Exposure of Workers to Pesticides, Arch. Environ. Health, 14, 622, 1967.

7) Wolfe H.R. et al., Health Hazards of the Pesticides Endrin and Dieldrin, Arch. Environ. Health, 6, 458, 1963.

8) Reary J.B., Monitoring of Exposure Hazards to Spray Operators and Villagers During Mosquito Trials with Bendiocarb (80W) in Thailand, August 1982, FBC Ltd., Internal Report RESID/80/89, 1981.

9) Reary J.B., Monitoring of Phenmedipham Contamination on Spray Operators During Low Volume Application of Betanal E in the U.K. 1980, FBC Ltd., Internal Report RESID/81/20, 1981.

10) Housden M.C. and Reary, J.B., Monitoring of Spray Operator Hazards During Trials with a CDA Application of Bendiocarb on Domestic Premises in Cambridge, 1980, FBC Ltd., Internal Report RESID/81/15, 1981.

11) Kangas J. et al., Exposure of Finnish Forestry Nursery Workers to Quintozine and Maneb, in Tordoir W.F., Van Heemstra E.A.H., (Eds), Field Worker Exposure During Pesticide Application, Elsevier, 1980.

12) Prinsen G.H., Van Sittert N.J., Exposure and Medical Monitoring Study of a New Pyrethroid after one season of spraying on cotton in the Ivory Coast, in Tordoir W.F. and Van Heemstra E.A., (Eds), Field Worker Exposure During Pesticide Application, Elsevier, 1980.

13) Wassermann M. et al., Toxic Hazards During DDT-and BHC-Spraying of Forests Against Lymantria Monacha, Arch. Ind. Health., 21, (6), 503-508, 1960.

14) Stearns C.R. et al., Concentration of Parathion Vapour in Groves After Spraying and Effects of the Vapour on Small Animals, Citrus Magazine, p22, 1951.

15) Culver D. et al., Studies of Human Exposure During Aerosol Application of Malathion and Chlorthion, Arch. Ind. Health, 13 (1), 37-50. 1956

16) Batchelor G.S., Walker K.C., Health hazards involved in the use of Parathion in fruit orchards in North Central Washington, Arch. Ind. Hyg. Occ. Med., 10, 522-529, 1954.

17) Batchelor G.S., Walker K.C., Elliott J.W., Dinitroorthocresol Exposure from Apple Thinning Sprays, Arch. Ind. Health, 13, 593-596, 1956.

18) Wolfe H.R. et al., Evaluation of the Health Hazards Involved in House-Spraying with DDT, Bull. W.H.O., 20, 1-14, 1959.

19) Lloyd G.A., Tweddle J.C., The Concentration of Dimefox in air Resulting from its Use on Hops, J. Sci. Food Agric., 15, 169-172, 1964.

20) Lloyd G.A., Bell, G.J., The Exposure of Agricultural Workers to Pesticides used in Granular Form, Ann. Occup. Hyg., 10, 97-104. 1967

21) Lloyd G.A., Bell, G.J. and Howgego, A.T., Contamination by Fenitrothion of the Aircraft Pilot, Humans at Ground Level, and Non-target Areas, in Holden A.W. and Bevan D., (eds), Aerial Application of Insecticide Against Pine Beauty Moth, p 63. 1978

22) Lloyd G.A., The Concentration of Bis-ethyl mercury phosphate Resulting from its Use in Glasshouses, J. Sci. Food Agric., 14, 845-848, 1963.

23) Comer S.W. et al., Exposure of Workers to Carbaryl., Bull. Environ. Contam. Toxicol., 13 (4), 385-391, 1975.

24) Fletcher T.E., Exposure to Dieldrin in Residual Spraying and the Fate of the Insecticide Absorbed by the Spraymen, W.H.O. Symposium on Pesticides, Brazzaville, Nov. 1959.

25) Jegier Z., Exposure to Guthion During Spraying and Formulating, Arch. Environ. Health, 8, 565-69, 1964.

26) Simpson G.R., Exposure to Orchard Pesticides, Arch. Environ. Health, 10, 884-5, 1965.

27) Simpson G.R., Beck A., Exposure to Parathion, Arch. Environ. Health, 11, 784-6, 1965.

28) Copplestone J.F. et al., Exposure to Pesticides in Agriculture, Bull. W.H.O., 54, 217-223, 1976.

29) Hayes A.L. et al., Assessment of Occupational Exposure to Organophosphates in Pest Control Operators, Amer. Ind. Hyg. Assoc. J., 41 (8), 568-75, 1980.

30) Kolmodin-Hedman B., et al., Field Application of Phenoxy acid Herbicides, in: Tordoir W.F., Van Heemstra E.A.H., (Editors), Field Worker Exposure During Pesticide Application, Elsevier, 1980.

31) Lavy T.L. et. al., Field Worker Exposure and Helicopter Spray Pattern of 2,4,5-T, Bull. Environ. Contam. Toxicol., 24, 90-96, 1980.

32) Lavy T.L. et. al., (2,4-Dichlorophenoxy)acetic Acid Exposure Received by Aerial Application Crews During Forest Spray Operations, J. Agric. Fd. Chem., 30, 375-381, 1982.

33) Dubelman S. et al, Operator Exposure Measurements during Application of the Herbicide Diallate, J. Agric. Fd. Chem, 30, 528-532, 1982.

34) Lavy T.L. et. al., Exposure Measurements of Applicators Spraying (2,4,5-Trichlorophenoxy)acetic Acid in the Forest, J. Agric. Fd. Chem., 28, 626-630, 1980.

35) Cohen B. et. al., Sources of Parathion Exposures for Israeli Aerial Spray Workers, Pesticides Monitoring Journal, 13 (3), 81-86, 1979.

36) Franklin C.A. et al., Correlation of Urinary Pesticide Metabolite Excretion with Estimated Dermal Contact in the Course of Occupational Exposure to Guthion, J. Tox. Environ. Health, 7, 715-731, 1981.

37) Libich S, et al, Occupational Exposure of Herbicide Applicators to Herbicides Used Along Electric Power Transmission Line Right of Way, Am. Ind. Hyg. Ass. J., 45, 56-62, 1984.

38) Stevens E.R., Davis, J.E., Potential Exposure of Workers During Seed Potato Treatment with Captan, Bull. Environ. Contam. Toxicol., 26, 681-688, 1981.

39) Atallah Y.H. et al., Exposure of Pesticide Applicators and Support personnel to O-ethyl-O-(4-nitrophenyl) phenyl-phosphonothioate (EPN), Arch. Environ. Contam. Tox., 11, 219-225, 1982.

40) Everhart L.P., Holt R.F., Potential Benlate Fungicide Exposure during Mixer/Loader Operations, Crop Harvest and Home Use, J. Agric Fd. Chem., 30, 222-227, 1982.

41) Gold R.E., Leavitt, J.R.C., Ballard, J., Effect of Spray and Paint-on Applications of a Slow Release Formulation of Chlorpyrifos on German Cockroach Control and Human Exposure, J. Econ. Ent., 74, 522-554, 1981.

42) Wright C.G., Leidy R.B., Dupree H.E., Insecticides in the Ambient Air of Rooms Following their Application for Control of Pests, Bull. Environ. Contam. Tox., 26, 548-553, 1981.

43) Hensley J.R., Supracide Worker Exposure and Dislodgeable Residue Studies in Alfalfa, Ciba-Geigy Corporation, unpublished data.

44) Hensley J.R., Igram Worker Exposure Study in Sorghum, Ciba-Geigy Corporation, unpublished data.

45) Hensley J.R., Nixon W.B., Supracide Worker Exposure and Dislodgeable Residue Studies in Citrus, Ciby-Geigy Corporation, unpublished data.

46) Kolmodin-Hedman B., Swensson A., Exposure of Workers Treating Conifer Plants with Lindane or DDT, Arh. Hig. Rada. Toksikol, 30 (suppl), 545-552, 1979.

47) Wojeck G.A. et al., Worker Exposure to Ethion in Florida Citrus, Arch. Environmental Contam. Toxicol. 10, 725-735, 1981.

48) Leavitt J.R.C., Exposure of Professional Pesticide Applicators to Carbaryl, Arch. Environ. Contam. Toxicol. 11, 57-62, 1982.

49) Chester G., Woollen B.H., Studies of the Occupational Exposure of Malaysian Plantation Workers to Paraquat, B.J. Ind. Med, 39, 23-33, 1982.

50) Maitlen J.C., et al., Workers in the Agricultural Environment: Dermal Exposure to Carbaryl, in Plimmer J.R., (ed), Pesticide Residues and exposure, ACS Symposium Series No. 182, ACS 1982, pp 83-104.

51) Belanger P., Health Hazard Evaluation Report NIOSH 80-024-783, 1980.

52) Carman G.E., et al., Pesticide Applicator Exposure to Insecticides During Treatment of Citrus Trees with Oscillating Boom and Airblast Units, Arch. Environ. Contam. Toxicol., 11, 651-9, 1982.

53) Draper W.H., Street J.C., Applicator Exposure to 2,4-D, Dicamba and a Dicamba Isomer, J. Environ. Sci. Health, B17(4), 321-39, 1982.

54) Housden M.C., Assessment of Operator Exposure and Deposits of Bendiocarb after ULV Application in Warehouses. FBC Limited Internal Report RESID/82/42, 1982.

55) Haverty M.I. et al., Drift and Worker Exposure Resulting from Two Methods of Applying Insecticides to Pine Bark, Bull. Environ. Contam. Toxicol., 30, 223-8, 1983.

56) Knaak J.B., et al., Safety Effectiveness of Closed-Transfer, Mixing-Loading and Application Equipment in Preventing Exposure to Pesticides, Arch. Environ. Contam. Toxicol., 9, 231-45, 1980.

57) Putnam A.R. et al., Exposure of Pesticide Applicators to Nitrofen: Influence of Formulation, Handling Systems and Protective Garments, J. Agric. Fd. Chem., 31, 645-50, 1983.

58) Staiff D.C. et al., Exposure to the Herbicide, Paraquat, Bull. Environ. Contam. Toxicol., 14(3), 334-40, 1975.

59) Wolfe H.R. et al., Exposure of Mosquito Control Workers to Fenthion, Mosquito News, 34(3), 263-7, 1974.

60) Kolmodin-Hedman B., Swensson A., Akerblom M., Occupational Exposure to Some Synthetic Pyrethroids (Permethrin and Fenvalerate), Arch. Toxicol., 50, 27-33, 1982.

61) Abbott I., et al, Spray Operator Safety Study. British Agrochemicals Association, London, 1984.

62) Grey W.E., Marthre D.E., Rogers S.S., Potential Exposure of Commercial Seed-treating Applicators to the Pesticides Carboxin-Thiram and Lindane, Bull. Environ. Contam. Toxicol., 31, 244-50, 1983.

63) Winterlin W.L., et al, Worker Reentry Studies for Captan Applied to Strawberries in California, J. Agric. Fd. Chem., 32, 664-72, 1984.

64) Davis J.E., et al, Potential Exposure to Diazinon During Yard Applications, Environmental Monitoring and Assessment, 3, 23-8, 1983.

65) Nigg H.N., Stamper J.H., Exposure of Spray Applicators and Mixer-loaders to Chlorobenzilate Miticide in Florida Citrus Groves, Arch. Environ. Contam. Toxicol., 12, 477-82, 1983.

66) Wojeck G.A., et al, Worker Exposure to Arsenic in Florida Grapefruit Spray Operations, Arch. Environ. Contam. Toxicol., 11, 661-7, 1982.

67) Kolmodin-Hedman B., et al, Studies on Phenoxy Acid Herbicides, I Field Study, Arch. Toxicol., 54, 257-65, 1983.

68) Longland R.C., Monitoring of Spray Operator Safety During Application of Prochloraz to Cereals in the UK, 1984. FBC Limited Internal Report RESID/84/63, 1984.

APPENDIX 2

WORLD HEALTH ORGANISATION

FIELD SURVEYS OF EXPOSURE TO PESTICIDES

STANDARD PROTOCOL

This is an abbreviated version of the WHO document VBC/82.1 published by Pesticide Development and Safe Use Unit, Division of Vector Biology and Control, WHO, Geneva. Permission to reproduce parts of the document is gratefully acknowledged.

1. GENERAL AIM

To examine the exposure of those workers involved in applying pesticides in order to be able to determine:

(a) whether the actual exposure constitutes a safe level;

(b) What protective measures need to be implemented to ensure safe use.

This can be achieved for a particular pesticide by the following activities:

(a) defining accurately the environmental conditions under which the pesticide is applied, the method and rate of application;

(b) collecting data on:

- (i) times of exposure;
- (ii) the quantity of pesticide applied by each man;
- (iii) details of protection used;
- (iv) the amount of pesticide coming into contact with the exterior of clothing, beneath clothing, and exposed skin;
- (v) quantitative tests of absorption and excretion of parent compound or metabolite;
- (vi) details of any effects of exposure.

If the exposure is equivalent to or exceeds a threshold level for safety, then the contributory factors involved (e.g. type of protection provided, personal hygiene, pattern of use) and their relative importance should be determined and appropriate recommendations for safe use should be made.

2. INDICATIONS OF CARRYING OUT A FIELD STUDY

Field studies should be carried out whenever the toxic properties of the substance and/or formulation together with the pattern of usage give cause for concern about workers' safety.

Considerations in determining this include:

(a) the acute dermal toxicity of the formulated product when dermal exposure is considered the most significant route of absorption;

(b) a significant opportunity for inhalational exposure at work of a material which is toxic by the respiratory route;

(c) where compounds have been found to cause adverse effects only at very high doses to ascertain that worker exposure will be at low levels, providing a significant safety margin;

(d) significant changes in formulation of compound or application procedures.

Studies may need to be repeated if there are changes in use patterns, application techniques or formulations which might alter the hazard to workers employed in application.

3. CHOICE OF SURVEY SITE AND GROUP

3.1 The spraying operation should be under single administrative control and should last a period appropriate to the pattern of use and the pharmacokinetics of the pesticide.

3.2 The pesticide should be applied in uniform concentration by uniform means throughout the operation.

3.3 The group studied should consist of four to 10 workers engaged in the application. If practicable, the occupational groups (mixers, loaders, sprayer, flagmen, etc.) should be considered separately. Ideally, the group should not have been exposed to the pesticide used during the survey or to any other pesticide with a similar mode of action during a relevant period preceding the commencement of the operation.

3.4 If it is necessary to select from spraymen, the selection should be on the basis of typical working clothing worn in the area in order that this might be standardised as far as possible. If possible, photographs of the men in their working clothing should be taken. If the workers engaged in the application are not experienced in pesticide application proper instructions and training should be given before the study commences.

3.5 Depending on the parameters measured and the period of the survey a suitable control group applying a blank formulation may be indicated.

4. SURVEY STAFF

4.1 Staff for the adequate supervision of the operational aspects, including proper application of the pesticide, have not been included as it is assumed that these would be provided for in a planned operation.

4.2 The minimum survey staff consists of a physician-toxicologist and one or two assistants of technician status.

4.3 The physician-toxicologist is responsible for:
(a) the general organisation of the survey;
(b) the keeping of records;
(c) medical supervision of applicators during the operation, including the collection and handling of specimens where necessary.
(d) the training of local staff in evaluation techniques.

5. METHODS

5.1 Cholinesterase determination

Organophosphorus compounds:

The principal effect of exposure to an organophosphorus compound is depression of cholinesterase. The degree of depression can be estimated in blood by, for example, the methods described by Michel,[1] Ellman,[2] or Voss & Sachsse[3] or other similar methods.

When field conditions would involve an interval between collection and analysis exceeding 10-12 hours, then it is necessary, even when samples are adequately cooled, to use a field kit such as the Tintometer method[4]. Cholinesterase test papers are also available; as they are of lower sensitivity, they are of more use for routine surveillance than for survey work.

Tests should be carried out on all personnel included within the study:
(a) once on each of three days preceding any exposure to

cholinesterase inhibitors; the mean of these three values provides the baseline for the individual against which subsequent values are matched;

(b) on working days when spraying has been completed and the spraymen have removed their protective clothing and washed or bathed;

(c) if a gap of one of more days occurs during the operations, a pre-work test should be carried out on the day that work is resumed;

(d) when any worker in the survey group complains of symptoms which could be related to pesticide exposure, a cholinesterase determination should be carried out immediately; thereafter cholinesterase should be measured at regular intervals until its recovery is established;

(e) if any worker has had an asymptomatic cholinesterase depression which has led to suspension from work, cholinesterase determination should be carried out on succeeding days until it is obvious that its recovery is taking place.

Blood should be taken from an uncontaminated area, preferably by the venous route. Capillary blood is usually used in field tests but in the case of direct acting cholinesterase inhibitors great care has to be taken to prevent the contamination from the skin. Venous blood may have to be used.

Insecticidal carbamates:

Due to the rapid reactivation of cholinesterase when inhibited by insecticidal carbamates, measurement of cholinesterase is not a feasible indicator of impending overexposure. In these surveys, in addition to other tests of exposure, surveillance should be adequate to detect any clinical signs.

5.2 Biological monitoring

5.2.1 Biological monitoring provides a quantitative measure of the absorbed pesticide resulting from exposure via all routes.

5.2.2 The objective of biological monitoring is to relate the level of parent compound or the specific representative metabolite in biological material to the total uptake. This implies that for this purpose specific and sensitive methods have been developed or should be developed to enable the determination of the above-mentioned substances in biological material.

In view of the limitations inherent to field conditions the biological material in which the pesticide or its metabolite is determined is usually blood or urine or both.

Since it may not always be possible to carry out the required analysis of the specimens close enough to the location of the field study, means of adequate preservation and despatching should be available.

5.2.3 Time and frequency of sampling and the volume of the specimen are dependent on the properties of the compound, the characteristics of the application procedure and the specific conditions of the study.

Detailed guidelines cannot be given; however, some general guidance is presented below:

(a) One or more pre-exposure samples must be analysed in order to establish a reliable baseline.
(b) During the application period an adequate number of samples should be collected in order to provide a good measure of the total uptake and an indication about possible transient retention.
(c) After the application period sampling should continue long enough to provide data on the elimination pattern.
(d) The decision to collect either spot samples or 24-hour urine samples in the case of monitoring urinary metabolites will depend on the properties of the pesticide, the level and the rate of absorption and the analytical requirements. A check on the adequacy of the 24-hour urine sampling should be carried out by, for example, measuring the creatinine content of that urine.
(e) Safeguards should be taken to avoid contamination of any sample with the parent compound.

5.2.4 The toxicological interpretation of the amounted determined in blood or in urine cannot be made without the knowledge of the pharmacokinetics, preferably in man.

5.3 Assessment of dermal exposure

This may be done by one of the two methods described below.

5.3.1 Disposable overalls and gauntlets

Each worker, who is to be assessed for dermal exposure, is required to wear a new disposable overall and gauntlets for a minimum period of one hour or more during any one day's spraying. If significant head exposure is likely, then a head pad as described below or a disposable hat should also be used. Care should be taken to ensure that the overall or gauntlets do not become saturated and should this occur a fresh overall and gauntlets must be worn. The exact periods of time and amount of pesticide used during the period that each overall(s)/gauntlets are worn must be accurately recorded.

On completion of each assessment period the overalls and gauntlets should be carefully taken off by an assistant to avoid cross-contamination. Each gauntlet should then be placed in a separate plastic or polythene bag and stored in a place out of direct sunlight and as cool as is practicable, prior to analysis.

The overall(s) is (are) immediately sectioned into the following parts:

(a) legs - above and below knee;
(b) arms - above and below elbow;
(c) torso - front and back.

Care should be taken to ensure that the instrument used for sectioning the overall is not contaminated. Decontamination can be carried out by swabbing with cotton wool soaked in an appropriate solvent. Each body part sample is then stored separately under identical conditions as are the gauntlets.

All bags should be labelled separately to include the following information:

(i) the worker's number;
(ii) number of day of survey - preceded by the letter D.

5.3.2 Exposure pads

Pads are prepared of α-cellulose, 10 x 10 cm backed with glassine paper or aluminium foil. Aluminium foil alone can be used with oily formulations. Benchkote® (Whatman) a white absorbent paper backed with polyethylene, has also been successfully used as an absorbent pad. Prior extraction of pad material is essential to ensure that there will be no interference with the analysis of residues of a particular pesticide (see section 7).

Pads should be fixed with masking tape covering the edges of the pad only, and on clothing with safety pins through the edge of the pad. On each day of application, four spraymen from the survey group should be chosen serially to wear exposure pads for at least one hour (as for overalls) except for the "skin pad" which should be worn the entire day. The attachment of the pads should be checked frequently and corrected if necessary.

Pads should be applied as shown below, on clothing (where worn) or on skin. This is set out for a right-handed man; if the man is left-handed opposites apply.

Arm: Upper surface of left forearm held with the elbow bent at right-angle across body, midway between elbow and wrist.

Leg: (1) front of left leg, above ankle;
(2) front of left leg, mid thigh

Trunk: (1) over sternum;
(2) on back between shoulder blade.

Head: If head not covered, forehead as high as possible to give good adhesion.
If head covered, on hat as close as practicable to top of forehead.

In addition, one pad (the "skin pad") is applied to the skin under clothing to the upper abdomen at approximately the same height as the arm pad and below the sternal pad but not shielded by it.

On removal at the end of the day, a 5 x 5 cm square should be cut from the pad. Care should be take that the scissors or other cutting instruments are not contaminated. Decontamination can be carried out as described above (5.3.1).

The central portion of the pad should be placed either in a bottle containing solvent or in a plastic envelope (see section 7). If a bottle is used, the cap of the bottle should have an aluminium liner inserted, the bottle closed and shaken gently to ensure that the whole pad is wetted with the solvent. The exterior of the envelope or bottle should be labelled clearly with:

(i) the number of the day, preceded by the letter D;
(ii) the worker's number;
(iii) the site of the pad, and whether it was attached to clothing "C" or skin "S" (e.g. "D3 6 Thigh C").

In addition, on two days during the trial period, a control to determine laboratory recovering rate of the pesticide should be carried out. To 5 x 5 cm squares cut from exposure pads or overalls, add carefully measured 0.05 and 0.1 ml samples of the pesticide concentrate and diluted spray respectively. These samples are then treated in the same way as those taken from the spraymen.

5.4 Respirators

Since the respiratory route of absorption is not significant in most standard pesticide applications, the measurement of respiratory exposure need not be carried out in each study. This is based on the fact that respiratory exposure does not usually exceed 1% of the dermal exposure, except possibly in the case of gaseous fumigants and exposure in confined spaces, or in cases where respiratory toxicity is very high in relation to percutaneous toxicity.

To estimate respiratory exposure, two of the men in each group of four wearing pads for the day should be asked to wear half-face cartridge-type respirators for the whole of the working day. Respirator pads should be fitted each morning on the external side of chemical filter cartridges; these pads are removed at the end of the day and placed entirely in a plastic envelope or bottle, clearly marked as in 5.3.2 above. The man should be carefully instructed on the care of the respirator in order to avoid any contamination at times when it is removed. All respirators should be carefully washed at the end of each working day, after the chemical filter cartridges have been removed.

If the men normally wear a surgical or gauge mask, respiratory exposure can be measured by fixing a pad to the underside of the mask in front of the nose. This is not as accurate as the wearing of respirators as set out above.

Individual respiratory samplers are also available and may be suitable for use in some studies.

6. RECORDS

6.1 Record of operation details

On the day spraying commences, Form 1 (Annex 2) should be completed. On each day subsequently, the data on the form should be reviewed and any changes recorded by completing a new form with details of the changes only.

6.2 Daily record and diary

Each day, from the day before spraying commences onwards, the daily record Form 2 (Annex 3) should be completed. The diary section of Form 2 should be used for recording any events which are relevant to the survey. Sickness in any man, whether or not he is included in the survey group, should be recorded, together with an account of action taken and result.

6.3 Daily personal record

Form 3 (Annex 4) should be started for each man in the survey group and should be entered daily.

7. LABORATORY PROCESSING OF EXPOSURE AND RESPIRATOR PADS AND OVERALLS

7.1 Before the survey

7.1.1 It is necessary to determine the recovery rate for the pesticide by the method used for its determination. In addition, it is necessary to estimate degradation of the pesticide between the time the exposure pad or overall is removed and the time of processing, and whether this can be reduced by the method of transportation. This is done by preparing patches of overall material or pads with known quantities of the pesticide and processing them immediately and after one, two and three weeks. If after three weeks recovery exceeds 75%, the overalls or pads can be transported in plastic envelopes, kept in the dark and as cool as practicable. Otherwise a suitable technique must be tested similarly. If solvents need to be used for transport and storage it is more practicable to use exposure pads.

7.1.2 A blank estimation using a sample of overall material or the pad only should be carried out to determine whether it is feasible to use an overall or whether prior extraction of pads is needed.

8. CALCULATIONS AND PRESENTATION OF RESULTS

8.1 Calculation of percentage toxic dose received by each man

8.1.1 Dermal exposure (from pads)

Exposed area:

This is calculated from the exposure pads, relating the quantity of pesticide in a pad of known area to the exposed area of the limb or part of the body as modified from Berkow (1931) Amer. J. Surg., 2, 315. These are added to give a total amount expressed in mg/day or per hour.

Head neck area = 1100 cm^2 reduce by a quarter
if hat worn = 825 cm^2........................head pad
Upper chest, V of neck = 150 cm^2...............sternal pad
Top of shoulders near neck = 300 cm^2...........head pad
Back just below neck = 100 cm^2.................back pad
Forearms = 1200 cm^2............................arm pad
Hands = 800 cm^2................................arm pad or hand wash
Legs from knees down = 2300 cm^2................lower leg pad
Upper legs from knees up = 3500 cm^2............thigh pad

These areas may need adjustment according to area unclothed.

Unexposed area:

Add areas of exposed parts and subtract from 18000 cm^2: the dermal contamination of the unexposed area is calculated from the "skin pad" underneath the clothing.

8.1.2 Dermal exposure (from overalls)

The potential dermal exposure is found by calculating the quantity of pesticide per section of overall and by adding the quantity of pesticide (in mg) of all sections of the overall covering parts of the body not normally covered by working clothes, plus the quantity of pesticide in the head pad (if worn) relating the mg/cm^2 to a figure of 1100 cm^2. If a head pad was not used the total exposure is body exposure + 10%. The total exposure is then related to the time the suit was worn, as follows:

$$\frac{\text{total quantity in mg} \times 60}{\text{total time (in minutes)}} = \text{total exposure (mg/hour)}$$

8.1.3 Respiratory exposure

If done, this is calculated and expressed in mg/day or mg/hour.

8.1.4 Percentage toxic dose per day or hour

This is calculated from these indices adapted from the method of Durham & Wolfe (1962) Bull. Wld. Hlth. Org., 26, 75-91, using the formula:

$$\frac{\begin{array}{c}\text{Dermal exposure} \\ \text{(mg/day or hour)}\end{array} + \begin{array}{c}\text{Respiratory exposure} \\ \text{(mg/day or hour)}\end{array} \times 10 \text{ (if measured)}}{\text{Dermal } LD_{50} \text{ mg/kg (rat)} \times 70/100}$$

8.2 Comparison of sites

All the results for a particular exposure pad site are grouped together and the mean and standard deviation (SD) are calculated. If any results fall outside the mean ± (SD x 3), these are excluded and a new calculation made of the mean and SD of those that remain. The purpose of this is to give an idea of the "normal" distribution after those results that represent some unusual contamination have been excluded. The chance of such exclusions being made in error is 1%.

REFERENCES

[1] Michel, H.O. (1949) J. Lab. Clin. Med., 34, 1564.
[2] Ellman, G.L. et al. (1961) Biochem. Pharmacol., 7, 88.
[3] Voss, G. & Sachsse, K. (1970) Toxicol. Appl. Pharmacol., 16, 764.
[4] Edson, E.F. (1958) World Crops, 10, 49; obtainable from the Tintometer Sales Co. Ltd., Salisbury, United Kingdom.

ANNEX 1

A list of metabolites and the pesticides with which they are related is shown in Annex 1 of the original WHO document VBC/82.1, not reproduced here.

ANNEX 2, 3 and 4

In the original WHO document VBC/82.1 sample record forms are provided for daily details of the spraying operation, the operational details, and personal practices for each individual in the respective annexes, not reproduced here.

INDEX